BRILLIANCE *and* FIRE

ALSO BY RACHELLE BERGSTEIN

———————————

*Women from the Ankle Down: The Story of Shoes
and How They Define Us*

HARPER

An Imprint of HarperCollins*Publishers*

BRILLIANCE

and

FIRE

A BIOGRAPHY
OF
DIAMONDS

RACHELLE
BERGSTEIN

HarperCollins books may be purchased for educational, business, or sales promotional use. For information, please e-mail the Special Markets Department at SPsales@harpercollins.com.

An extension of this copyright page, "Photographic Sources," appears on pages 373–75.

FIRST EDITION

Designed by William Ruoto

Library of Congress Cataloging-in-Publication Data has been applied for.

ISBN: 978-0-06-232377-4

16 17 18 19 20   OV/RRD   10 9 8 7 6 5 4 3 2 1

For Andrew

_____

Contents

Preface   xi

1   THE BUYERS
How the American Rich Came to Love Diamonds   1

2   THE SEEKERS
How One Sickly Englishman Took Over the
Diamond Industry   19

3   THE LOVERS
How Young Women Came to Expect Diamond Rings   41
(Part I)

4   THE SHOW-OFFS
How a Single Stone Bewitched Its Owners—   57
and Captivated America

5   THE OPTIMISTS
How the Industry Survived Two World Wars and   79
the Great Depression

6   THE SELLERS
How Young Women Came to Expect Diamond Rings   99
(Part II)

7   THE QUEENS
How Wallis Simpson Earned Her Diamonds   117

8   THE STARS
    How an Advertising Agency Got Diamonds Ready   135
    for Their Close-Up

9   THE WINNERS
    How Elizabeth Taylor Learned to Speak Italian   157

10 THE MEDDLERS
    How Little Stones Got Big   173

11 THE INVENTORS
    How to Make a Diamond (and Provoke De Beers)   193

12 THE ILLUSIONISTS
    How the Magicians Protected Their Tricks   213

13 THE MASTERS
    How the Australians Democratized Diamonds   229

14 THE STUNNERS
    How Tough Guys Started Wearing Diamonds   247

15 THE CRITICS
    How a Global Crisis Changed the Meaning of "Forever"   267

16 THE INNOVATORS
    How to Sell Diamonds in a (More) Enlightened World   293

    Acknowledgments   319

    Notes   323

    Index   353

Preface

Her mornings were always the same. For hours she might sit at the kitchen table with a cup of tea, wearing casual slacks and a blouse, scanning the obituaries or the gossip columns as the sun lifted toward noon and light flooded through the windows, reflecting on the 4.82-carat marquise-cut diamond ring on her finger. She was a housewife in her seventies with five grown children, and she was, understandably, bored. Occasionally she had visitors; some nights she went out to dinner with her husband, to one of the handful of local restaurants where they were regulars. He always wanted to go somewhere familiar, where the staff greeted him by name, and she rarely argued—if she wished they'd try a new place, she didn't say, and deep down she appreciated the extra fuss the waiters made over her.

But whether or not she had anywhere to go, she wore her diamonds every day.

They were first-generation children of poor European immigrants: she'd grown up on a farm in northern New Jersey just a few miles away while her husband learned construction from his father. By the look of things, they lived out their American dream. They got married on the eve of World War II, and when he returned home safely from a tour with the navy, she loaned him the money to buy his first truck, which he—fiercely determined to succeed and motivated, if not exactly championed, by her appetite for the finer things—turned into a fleet. He made a small fortune in roofing, enough to support a family of seven and to buy a hulking colonial house on a hill. As the age of American consumerism unfolded, they filled it with expensive things: matching Cadillacs in the garage, custom suits and dresses, fur coats and the kind of dazzling jewelry she saw beautiful Hollywood actresses wearing in magazines. She loved Zsa Zsa Gabor, Elizabeth Taylor—the famous women who always looked glamorous and who, just by being themselves, promoted a very specific kind of midcentury materialism.

Still, they were unhappy. He drank too much; they both screamed. Like so many suburban couples in the 1950s and 1960s, they were deeply committed to the dual occupations of breeding and acquiring, and the resentment began creeping in when they both realized, independently, that neither vocation was entirely satisfying. They took out their frustrations on their kids, and on each other. But no matter how bad things got, every year—for her birthday, or Christmas, or their anniversary—he bought her diamonds. And every year, as she opened the little box and onlookers fawned, she gasped with pleasure but then remembered herself, and glibly pronounced the stones too small. Perhaps it was her way of telling him that even a lavish gift—whether it was a stunning tennis bracelet or a new icy-white pair of matched studs—wasn't enough to make up for the way he behaved the rest of the time. Or perhaps it went deeper than that. By admitting that she loved the presents, and thus allowing herself to let go of her long history of middle-class striving, she'd be relinquishing the very thing that set her alight—the desire to be richer, fancier, better.

He retaliated in his own way; he kept buying her bigger diamonds, but of lesser and lesser quality. If she noticed the scheme, she didn't say. And he probably never saw the way that she admired her baubles when she thought no one was looking. Sitting at her kitchen table, her jewels shimmering in the natural light, my grandmother—the Elizabeth Taylor of northern New Jersey, the woman I've known to love diamonds more than anyone else in my life—often looked down at her hands and smiled.

⬦⬦⬦⬦⬦⬦⬦⬦⬦⬦⬦⬦⬦⬦⬦⬦⬦⬦⬦⬦⬦⬦⬦⬦⬦⬦⬦⬦⬦⬦⬦

What is it that makes the glittering rocks so captivating? When I set out to write this book, I wanted to trace the role of the diamond in our culture, paying special attention to the fundamental complexities of the stone as a symbol, simultaneously, of magnificent success and romance—but also of shameful avarice. Although I admit I find diamonds quite beautiful, and at times even mesmerizing, I don't yearn for them the way my grandmother did—my tastes are a bit simpler, and I don't possess that particular strain of restlessness I've come to associate with the most enthusiastic and prodigal gemstone buyers. Even so, it probably isn't a fair comparison. During my grandmother's time, diamonds spelled royalty and riches and undeniable accomplishment—there were many people behind the scenes of the industry working around the clock to make sure of that. Today the stones mean all of those things, but they're also scarred by other, nastier associations that have surfaced in recent years: geologic devastation, slave labor, disrupted indigenous populations, unthinkable violence. It's no wonder that even though diamonds are still in fashion, it's not terribly fashionable to like diamonds.

Those realities inform *Brilliance and Fire*. The stones are literally, and also figuratively, surprisingly multifaceted: they've meant many different things to different people through the years. Once upon a time, they were believed to stave off ghosts, to avert death (if the diamond was clear), to invite death (if it wasn't), to promote fertility, to rupture the innards if ingested, and to prompt reconciliation between couples—just to name a few of the otherworldly powers that have been ascribed to them. To the marginalized European Jews in the Middle Ages, they meant the possibility of employment; to their descendants during the Holocaust, they provided financial—and portable—means for escape. To the Indian maharajas in the seventeenth century who treasured diamonds, they were a source of mystical strength. To the wealthiest residents of New York City during the Gilded Age, they offered proof of relevance on the global stage, as well as a means for upper-crust social acceptance.

Knowing this, we may be amazed to think that the rocks come out of the ground looking—at least by comparison with the finished product—fairly unspectacular. A typical rough diamond is irregularly shaped and has a milky, soapy white or even yellowish hue; the very first person to suggest cutting and polishing one to sparkly perfection deserves a long entry in the annals of lost history, perhaps alongside the culinary pioneer who decided to try eating a lobster. And like those ancient crustaceans, diamonds are products of another age: they form deep in the earth's mantle under specific conditions of very high pressure and temperature, and then shoot toward the surface by way of subterranean volcanic eruptions. Scientifically speaking, they're quite special. They're carbon—just like coal, and graphite, and you and I—but in the case of the precious rock, the element forms in the shape of a sturdy crystal lattice that accounts for its unique and most covetable properties. Namely, diamonds are hard—they hold the title of hardest naturally occurring substance known to man—and they're also chemically stable, very good thermal conductors, and, to use the proper scientific term, optically dispersive. It is this final attribute that leads admirers like me to pause at the Tiffany & Co. windows: diamonds catch the light. When a diamond cutter approaches a rough stone, he has many goals in mind, but the primary one is usually maximizing the particular specimen's reflectivity, which largely contributes to its most abstract yet desirable quality: the diamond's undeniable, almost visceral beauty.

The jeweler's job is to highlight that beauty and, ideally, to enhance it, whether a stone is a quarter carat, a little over an eighth of an inch (4.1 millimeters, mm); or a carat, just past a quarter of an inch (6.5 mm); or an enormous three carats, measuring around three-eighths of an inch (9.4 mm), which still sounds tiny in theory until it's compared with the diameter of an average woman's ring finger (slightly smaller than 16.5 mm, the measurement for a standard size-6 ring). Since diamonds were first discovered up until the present day, craftsmen have worked to innovate the process of gem fashioning, giving us timeless cuts like the round brilliant, emerald, and pear shapes, as well as intermittently trendy ones like the princess cut, usually square with sharp corners; the marquise, which is oval like a football; and the romantic heart shape. The world's finest jewels are indisputable works of art, created by people who are as passionate and committed to their craft as master painters, architects, authors, poets, and filmmakers all are. Even if you're not inclined to fawn over baubles, it's difficult to look at Jacob & Co.'s 2015 Billionaire watch, for instance, composed of 260 carats of invisibly set, matched, emerald-cut, winking white diamonds,

and not feel a certain awe, akin to the experience of reading a great
work of fiction or staring up at a towering skyscraper. It's respect
for the vastness of human accomplishment and even, perhaps, the
stirrings of the sublime.

That said, it's also easy to understand how someone could look
at the same piece—designed by the man once known as "Jacob
the Jeweler" along with notorious Italian businessman Flavio
Briatore—and have an entirely different reaction, disgust with
the excesses it represents and, especially, with its mind-boggling
$18 million price tag. Again: that's the fundamental paradox of
diamonds. For those of us who would like to believe the world has
changed since the nineteenth century, when Fabergé designed
elaborate gem-studded eggs for Russian tsars while the peasants
starved, a piece like the Billionaire watch can be unsettling, to say
the least. But again and again, diamonds—and the multilimbed
industry that exists to unearth, distribute, and promote them—
have proved their resilience. Even at the turn of the millennium,
when the "conflict diamond" crisis was so well publicized that it
became the subject of a big-budget Hollywood movie, ordinary
people—in the West, but also in the East—didn't stop buying the
stones. It's safe to say that we're drawn to them. There's something about diamonds that compels us to indulge, and also to
forget, the way I did when my husband presented me with an antique diamond-and-sapphire engagement ring and my eyes filled
with tears, and the way I suspect my grandmother did when she
sat by herself, watching the light dance.

To me, the fact that a colorless stone can be all of these things—
among them, a vital prop in a ritual that many consider to be the
ultimate gesture of love and also our clearest emblem of ugly
materialism—is what makes the diamond so profoundly fascinat-

ing. This book attempts to gaze straight into the diamond's many, many faces. It also tells a unique and, as I've come to believe, very revealing story: about how we as a society have come to cherish an unusual rock, and about the individuals who have made that rock their lives.

BRILLIANCE *and* FIRE

I

The Buyers

HOW THE AMERICAN RICH
CAME TO LOVE DIAMONDS

New York

I n the dark winter days of early February 1897, New York City
was uncharacteristically lively. A fancy costume ball, to be held
at the Waldorf hotel on the tenth of the month, had so absorbed
the imagination of the public that it seemed there was nothing else in the world anyone wanted to discuss. For the party's
1,200 invitees, representing the pinnacle of New York's high society, the evening provided a chance to savor the advantages of wealth
and to fortify a position within the elite. For everyone else—the
journalists, shopkeepers, and ordinary folks—the Bradley-Martin
ball was a controversy worth debating. Was it a cheerful distraction
from hard economic times? A citywide boon? A sign of the increasing self-indulgence of the rich? Or a full-fledged abomination? For

weeks, as would-be guests visited their dressmakers and pulled dusty family heirlooms from their safes, the rest of the city—and the world—waited for the who's who of New York to alight from their carriages on Fifth Avenue and make their debut as the new American gentry.

The country was in the throes of a terrible depression. Given the dismal state of things, one local preacher, the Reverend Dr. William Rainsford, had gone so far as to counsel his largely well-off parishioners at St. George's Episcopal Church to forgo attending entirely. But in spite of the hubbub, the ball's hostess, Cornelia Bradley-Martin, had no second thoughts about her upcoming namesake evening. A seasoned party giver, she had thrown two hugely successful events with her husband over the past few years: dinner for three hundred at Delmonico's followed by a cotillion in 1880 and a full-scale ball at their home five years later, which was so well received it was said to rival parties put on by the Vanderbilts. Yes, Mrs. Bradley-Martin understood that this ball was different. The guest list was enormous, and in New York City, where the "panic" began in 1893, unemployment was still so high that former hardworking people were reduced to indigence, waiting in breadlines and begging. Even the wealthiest citizens felt the effects of belt-tightening. But that was precisely the point! Couldn't Reverend What's-His-Name see that?

Nearly a year prior, Bradley-Martin was having breakfast in her chic double residence at Twenty and Twenty-Two West Twentieth Street when the idea came to her. She had the morning paper spread open and was so sickened by reports about the continuing misfortunes of the poor that she was barely able to eat. Suddenly, it hit her: she would host a ball! And not just any ball—the biggest, most lavish event the city had ever seen, so appealingly over-the-top that even the most reclusive and tightfisted members of society would be compelled to attend. It would be a costume party; that way, guests would be required to visit milliners, dressmakers, wig makers, and jewelers in preparation for the ball rather than pulling some long-forgotten gown from the back of the closet and passing it off as new. That was the genius of the plan—she wanted New Yorkers to shop. The heavy economic crisis refused to lift, and this was her version of a stimulus: dinner and dancing and dollars exchanged by way of Cornelia Bradley-Martin.

Her husband approved, and she immediately started planning, booking the Waldorf hotel on Fifth Avenue and Thirty-Third Street for the event, even though balls were customarily held in the host's expansive home. A theme was chosen: the court of Louis XIV. It would flatter the guests, who, in the absence of American royalty, considered themselves the next best thing. Bradley-Martin worked with the Waldorf's eccentric and much adored caterer, Oscar, on the menu; collaborated with florists to flood the hotel with roses, clematis, and orchids; and dreamed up a vision to transform the grand ballroom into the Great Hall of Mirrors at Versailles. The Bradley-Martin ball would be a party worthy of the Vanderbilts, Havemeyers, Stuyvesants, and Astors, the old guard of the American elite. It would disburse hundreds of thousands of dollars across food purveyors, artisans, and waiters, all of whom had suffered too long from lean times. The chilly night of February 10, 1897, would make history.

Or, at the very least, it was an excuse for Mrs. Cornelia Bradley-Martin to wear her diamonds.

◇◇◇◇◇◇◇◇◇◇◇◇◇◇◇◇◇◇◇◇◇◇◇◇◇◇◇◇◇◇◇◇◇◇

The Gilded Age was the era of the puffed-up robber baron and his socialite wife, a time when the vast undertaking of a cross-country railroad forever altered the American landscape. It was also, on the other hand, a historical moment defined by grave instability; the very panic of 1893 that the Bradley-Martin ball was intended to address was brought on in part because the railroads were built too quickly to recoup expenses, resulting in bankruptcies, unemployment, and widespread poverty among those who had proudly considered knitting the tracks together to be their livelihood. In Manhattan, which had only recently surpassed Boston and Philadelphia as the center of American commerce, the era's greatest profiteers and its biggest casualties lived alongside one another on a fourteen-mile stretch of island. The winners' spoils couldn't have been more ostentatious: mansions, carriages, art, furs, silver, gold, and, of course, diamonds.

The wealthy considered diamonds to be a good investment, but the sparkling stones were prized mostly for their other, more intoxicating charms. Recently, in 1897, a new source of diamonds had been discovered in South Africa, but for hundreds of years they had been found only in a handful of exotic, far-off locations. The most fruitful of these were India and Brazil, where rough gemstones lay hidden in riverbeds and were traded in diamond cities—the most famous being Golconda, India—across the world. Because diamonds were relatively rare and indigenous to regions of the world where no typical Western men traveled, they were prohibitively expensive, and thus appealed to the royals and aristocrats who could afford them.

For the American rich, that diamonds were strongly associated with European royalty only enhanced the stones' natural allure. At the turn of the century, few men wore diamonds, and so it was the responsibility—and pleasure—of the wife to physically reflect the couple's affluence. The job proved mesmerizing; in the glittering rock, an American upstart saw her legacy refracted so that it recalled the great princesses and queens of history. This was especially tempting to a woman like Cornelia Bradley-Martin, who, despite her family's wealth, felt insecure about her place within New York's elite.

Cornelia was hardly a queen: she had made a humble marriage to a man from Albany, New York, who spotted her, a blond, pink-cheeked bridesmaid, at a friend's wedding. As newlyweds, they'd moved in with her parents. It was only after her father, a retired merchant who was rumored to have money but lived modestly, passed away that Cornelia and her husband, Bradley Martin, were able to fulfill their true social ambitions. To their shock, her father had saved over $5 million, which he willed almost entirely to his daughter.

With their windfall the couple took an extended trip to Europe, where Cornelia Martin, much to the amusement of her higher-born friends, began hyphenating her last name in the style of the English aristocracy. And diamonds—they were the passport to another stratum. In more ways than one, gemstones connected the American rich to older, more established cultures across the ocean. The best-quality ones had long been found in Europe, in fashionable cities like Paris, where the top jewelers like Cartier, Vever, and Boucheron all had their shops. Women like Arabella Huntington—widow of railroad magnate Collis Huntington, born humbly but possessed of an estimated $70 million (almost $1.9 billion today) after her husband left his fortune all to her—made a practice of traveling to the Place Vendôme and the Rue de la Paix to buy clothing, art, and jewels. The experience of shopping there was as luxurious as the merchandise. Pausing to pet the owner's cat Wladimir at the Boucheron boutique, making conversation *en français* with a fawning clerk while he handwrote the receipt in careful script—

these were the experiences that distinguished the sophisticates. In February 1902, the *Los Angeles Times* reported Huntington's return stateside through customs, where she declared $75,000 (about $2 million in 2014) in purchases: the highest number in the history of the Port of New York. Like Cornelia Bradley-Martin, she felt self-conscious about her unprivileged upbringing, and believed it was worth the 60 percent importation tariff at customs to be able to wear the most fashionable items.

New York City boasted its own tucked-away jewelry district, adjacent to the seaport on the short downtown street of Maiden Lane—but it couldn't compete with the glamour of Paris. Still, one homegrown American jeweler was doing its best to deliver a posh, inimitable experience to its customers. Tiffany & Young opened its doors at 259 Broadway on the morning of September 18, 1837. Owned by two grade-school friends—later doubly linked by Charles Lewis Tiffany's marriage to John Burnett Young's sister Harriet—it began as a "fancy goods store" that offered only a small selection of jewelry and specialized in paper goods and curiosities: stylish little indulgences like umbrellas, walking sticks, porcelain vases, chinoiserie, and fans. The first day of business, the store took in just $4.98 ($119 today). Tiffany and Young second-guessed the location—it was risky being so far uptown of Maiden Lane—but the presence of Alexander Turney (A. T.) Stewart, who ran the store two doors down and would become the most powerful dry goods seller in the nation, reassured them.

Sure enough, business improved steadily, and by the holidays, the numbers were encouraging. Then, one year later, disaster struck: a New Year's Day burglary nearly wiped them out. The $4,000 (almost $100,000 today) in losses hurt, but Charles L. Tiffany, who kept Tiffany & Young's books, refused to be driven out of the city quite so easily.

At twenty-five, the handsome clerk with a long, straight-arrow nose and a chinstrap beard had a surprising amount of experience under his belt. Living in Connecticut, Charles's father, Comfort Tiffany, a cotton manufacturer, recognized his eldest son's talents and offered him free rein at a country store he opened on the Quinebaug River, where the senior Tiffany's new mill was under construction. Just fifteen years old, Charles excelled, and when the time came for Comfort to rename his expanding manufacturing company, he opted for C. Tiffany & Son.

Charles was represented in name only. Much to his father's disappointment, Charles L. Tiffany had no interest in following him into the cotton business. He worried that the product would eventually diminish in value and, more, wanted to succeed in a place like New York, where ambition either flourished or suffered a swift demise. Thankfully, for him it was the former. Tiffany & Young survived the robbery, and by 1841, the firm was well established within the world of New York City retail.

But there was always room for improvement. For years, Charles L. Tiffany had made a practice of walking down to the damp, bustling wharves and buying imported products straight off the ships, products that had more than satisfied the customers' taste for the international. However, greater success for the store meant a savvier clientele, and with that, Tiffany and Young understood that having limited contacts in Europe was holding them back. They joined up with the wealthy Jabez Lewis Ellis, who was able to finance buying trips overseas, and who had the connections in England, Germany, Italy, and France to make them worthwhile. The three became Tiffany, Young & Ellis, and by 1845, they were able to advertise their selection of "French Jewelry: at the particular request of many who are accustomed to buy them, T. Y. & E. have determined to pay special attention to this branch of their business, and will receive by the French Steamers and packets, a limited number of every *new style*." It worked: just a year later, they expanded to a larger storefront nearby, funded by a thriving city's appetite for luxury goods like watches, silverware, and jewelry. To whet that appetite, the company opened a silver workshop, beginning its crucial transition from retailer to jeweler. It printed a "Catalogue of Useful and Fancy Articles": a blue booklet slightly larger than a woman's palm, small enough to fit in a jacket pocket. With that, Tiffany, Young & Ellis had a signature color—somewhere in between pale peacock blue and aquamarine—and embraced it. The iconic little blue box was introduced in 1853.

That same year, John B. Young and J. L. Ellis both retired, and Charles Lewis Tiffany, at forty-one sporting a full beard still more dark than gray, made a decision. He was a sturdy patriarch—the father of five children, with one more dead and buried—and he'd consistently made choices that kept the company on top. He was as reliable as he was fiercely determined; he walked to work every morning in his top hat and Victorian tailcoat, arriving at the store at nine sharp, and stayed late, costing him precious time away from his family. In his entire career, he'd never once stayed home sick. He'd designed beautiful store windows that attracted foot traffic. He'd insisted that all of the merchandise have a set and visible price, at a time when haggling with customers was the status quo. Perhaps most important, when John Young arrived in France in 1848 on a buying trip and found the city in chaos, with the Orléans monarchy overthrown and the aristocrats looking to sell off their diamonds for a song, Tiffany counseled him to bring home as much diamond jewelry as he possibly could, even if it meant tying up most of the company's assets. Subsequently, as the store sold off the spoils, the press started referring to Charles L. Tiffany as the "King of Diamonds." Absent his partners, Tiffany felt he had earned the right to change the company's name—and so, Tiffany & Co. was reborn.

Tiffany & Co. continued to thrive, even during the unlikely Civil War period, when demand for luxury goods plummeted; Tiffany reacted by stocking the shelves with military supplies, uniforms, and ornamental swords. The next great success came in 1867, when Tiffany & Co.'s sleek American silver won third place at the Paris Exhibition, shocking the international community by beating out older, establishment jewelers. It followed up with another win for its silver at the Centennial Exhibition in Philadelphia in 1876 and a gold medal in jewelry at the Paris Exhibition of 1878—a real coup with the French on their home turf. With Tiffany & Co. branches operating in Paris and London, and a new, six-story flagship at New York's Union Square, the company, under Charles Lewis Tiffany's guardianship, was making its name as the first competitive American jeweler, and one that ranked, officially, among the best in the world.

In reputation, Tiffany & Co. was gaining on its European competitors and challenging the supposition that only time-honored companies knew their way around baubles. Yet the truth was that America, and therefore any American jeweler, was at a disadvantage when it came to diamonds. Many older countries had monarchs who collected diamonds, as well as rubies, emeralds, sapphires, and pearls, and so the largest known and most historically significant stones all remained overseas. The South African mines provided a steady stream of new, smaller diamonds, but with the exception of a diamond girdle, which was rumored to have belonged to Marie Antoinette, and which John B. Young brought home with his loot from France in 1848, Tiffany & Co. had no special access to large or historic stones, as opposed to houses, like Bapst (in France) and Garrard (in England), that had developed long-standing, privileged relationships with royalty.

Charles Lewis Tiffany likely had this in mind when he purchased a 287.42-carat South African rough yellow diamond in 1878 for $18,000 ($414,000 today). When cut down to less than half its weight—128.54 carats—it was still one of the largest known canary diamonds, and in size rivaled the fabled Koh-i-Noor. The Tiffany Diamond would become the house's showstopping namesake gemstone, a glittering feather in Tiffany's cap—but it was still just one jewel. Which is why Charles Tiffany couldn't resist when he heard the surprising news that in May 1887, the French government would be hosting a liquidation sale at the Pavillon de Flore at the Tuileries. The draw? Hundreds of years of Gallic avarice, indulgence, exploration, and acquisition. The auction was called Les Diamants de la Couronne de France: The Diamonds of the French Crown.

The event was politically motivated, and as with most of France's politics in recent history, the decision-making process leading up to it was tempestuous. The Third Republic government, which came to power with the fall of Napoleon III, was making a statement: the French royals were gone for good. The Bonapartists, the Orleanists, the Bourbons—those who wished to restore the monarch—considered the sale a travesty. Some jewelers, like Germain Bapst, who was descended from a line of crown jewelers and who personally created many custom pieces for the Empress Eugénie, was devastated, and quite vocal about the breadth of loss that the sale represented to him. Others, like Gérard Boucheron, jumped at the chance to get his hands on stones that would have otherwise been elusive to a person of his experience; though he was already quite respected, with his shop open twenty-four years, compared with someone like Bapst he was still green. The pieces on offer were in various states of repair; some were left untouched since the days they'd decorated the necks, wrists, and fingers of kings and queens, while others had been broken up to get at the stones: the equivalent of scrapping a car for its parts. But among the loose diamonds and anonymous lots on offer (among them lot 25, "A Corsage Bouquet—Two thousand six hundred and thirty-seven brilliants, 132 5-16 carats; 860 rose diamonds"; and lot 27, "Diadem, Emeralds, and Brilliants—One thousand and thirty-one brilliants, 1,076 carats; 40 emeralds, 77 carats") were some truly extraordinary gems. Seven Mazarins—large, beautifully cut diamonds, ranging from sixteen to nearly twenty-nine carats, said to have been acquired, or recut, under the supervision of the cardinal of the same name—were being advertised. It was rumored that the French government would be selling the Regent Diamond, a huge, 140.64-carat, colorless square-cut stone; but under so much scrutiny already, the government reneged, hoping to pacify the critics by agreeing to reserve a few matchless items for placement in the national museums.

The afternoon of May 12, 1887, six hundred dealers and spectators descended on the Pavillon de Flore, greeted by a nearly equivalent number of policemen. Tiffany & Co.'s representative was there, as were Bapst and Boucheron—but, according to the *New York Times* correspondent, the clamor leading up to the sale proved not to be predictive when it came to the level of enthusiasm exhibited by the French. "The sale of the Crown jewels seems to have excited more interest in America that it does here, where little is felt and still less manifested among the natives," he wrote. One by one, the first ten lots—the sale would last eleven days total—were escorted around the room. The bidding was polite. Tiffany & Co. purchased lot 10, a cascading four-strand necklace, each strand made up of rows of brilliants, for 183,000 francs. Another big winner was Belgian jeweler Alfred Doutrelon, but as the *Times* reporter pointed out, buying the real pieces came with no guarantee of exclusivity. "Of one thing we may be certain, that six months hence there will be ten times as many Crown jewels worn by individuals as were ever in the royal collection. . . . Such houses as that of Tiffany are above suspicion, but there are many unprincipled peddlers who will profit by popular credulity to pass off the spurious for the genuine." Indeed, a few months before the sale, the government caved in under public pressure to document the collection and brought in a photographer, whose images then became a catalogue, which was widely distributed for publicity. With these photos in hand, some less scrupulous jewelers had already gotten to work copying the items—not to sell them off as genuine, necessarily, but to offer bargain buyers their own scintillating brush with royalty.

In the days ahead, Tiffany & Co. surprised competitors by buying far more than was expected, outbidding establishment jewelers like Bapst and driving up prices. The bidding, which during the first few days had been restrained, grew more feverish and aggressive. Ultimately, Tiffany & Co. walked away with a stunning one-third of the lots. Among its wins—a *sévigné* (bow) brooch made up of 321 brilliants and a corsage of diamonds broken up into thirteen smaller lots—were four of the Mazarins, bringing its total eleven-day expenditure, after commissions, to $487,956 (more than $11.8 million today). Tiffany & Co.'s representative insisted to the press that, despite the historic nature of the diamonds, he did not get starry-eyed and overpay: "In making our purchases, we took care to pay nothing for the historical association. The trade value of the jewels was the sole guiding consideration." Maybe so, but Charles Tiffany and his colleagues understood that even if they were unmoved by the romantic lineage of the stones, the American tycoons and their socialite wives waiting in the shadows wouldn't be quite so levelheaded. And who were these mystery buyers, many of whom were vacationing on the Continent and took items off Tiffany & Co.'s hands so quickly that they never had to be shipped back to the States? Again, the company's buyer was businesslike: "I know it would be an interesting piece of news to let you know the names of the Americans who have secured some of the French Crown jewels from our house, but it is our rule never to supply such information."

No matter—those in the know kept their eyes peeled at the upcoming society affairs. Mrs. Joseph Pulitzer blew her cover when she wore lot 10—the four-strand diamond necklace—to a ball in Paris soon after the sale. And what about the dangling corsage brooch with the 20.03-carat centerpiece stone, the two ruby-and-diamond bracelets once owned by the Empress Eugénie, and the stately gem cluster that had belonged to a larger design (a girdle) but was still gorgeous in its own right? What lucky woman got to add those pieces to her private collection?

Cornelia Bradley-Martin seized every opportunity to wear them.

◇◇◇◇◇◇◇◇◇◇◇◇◇◇◇◇◇◇◇◇◇◇◇◇◇◇◇◇◇◇◇◇◇◇

At last, the evening of the ball arrived. Amid all the debate, criticism, obsessive reporting, and even outright mockery, Cornelia Bradley-Martin kept focused on the planning, likely feeling just the slightest bit smug that the Metropolitan Opera House was to debut its production of the popular *Martha* that same night and, alas, the opening would be poorly attended because the crème of society would be enjoying her party instead. She'd received over one thousand positive RSVPs, and many of the city's social leaders were throwing costume-themed dinner parties preceding the ball so that attendees could eat without worrying about a wardrobe change afterward.

Of course, some other partygoers had rented rooms in the Waldorf so that they could get dressed at their leisure and avoid the line of onlookers, reporters, and policemen who had gathered in front of the hotel. Anticipating a scene, Bradley-Martin had arranged for guests to enter through the proprietor's private entrance on West Thirty-Third Street, where a doorman diligently collected invitations and kept his eyes peeled for gate-crashers. People began arriving around ten p.m. Once inside, they were whisked into dressing rooms, where hired professionals fixed any errant hairs or lopsided headpieces resulting from wintertime travel in drafty open carriages. Gathering in the smoking chamber, the men compared swords and chuckled nervously about their short pants and tights—an immodest-seeming style that had gone out of fashion by 1897 but, in the interest of authenticity, was making a gallant return for one night only.

Slowly, guests were escorted into the reception room, where Mr. and Mrs. Bradley Martin sat poised on a dais to greet them, and then finally toward the grand ballroom: a stunning, high-ceilinged expanse of coppery greens and reds, accented with portraiture, hidden balconies, and Corinthian columns. The orchestra played Liszt, Franck, and Wagner, and behind the scenes, cooks hastened to prepare a lavish meal. After the quadrilles—by then, an already old-fashioned folk dance performed by four honored couples at a time—supper was served at midnight, a mix of hot and cold dishes including oysters, chicken with truffles, foie gras, beef jardinière, and ham, followed by dessert. Food was made available all night so guests could take a break from dancing to replenish. Little rooms around the hotel had been staged so that someone looking for a few moments' respite could wander through the premises with his glass of Moët & Chandon without feeling he had accidentally let the gates of Versailles lock behind him.

The costumes were a bit more varied than the theme of Louis XIV might have suggested, but Cornelia Bradley-Martin was satisfied on two counts: One, the guests had gotten into the spirit and transformed themselves in one way or another. Two, hers was by far the most extravagant outfit of the night. Elisha Dyer Jr., the Rhode Island politician who led the evening's cotillion, was dressed as Francis I in rich purple velvet embroidered in gold. Mrs. Astor was guided by historical fashion rather than a specific character, and wore a blue velvet dress trimmed with fur and lace. Her husband, John Jacob Astor, in purple satin, was Henry of Navarre. Mrs. Herbert Pell, mother of the future US congressman, turned heads as Catherine of Russia. There was a Pocahontas (J. P. Morgan's youngest daughter) and a shepherdess, but there was also a Cardinal Richelieu as well as assorted Marie Antoinettes. Bradley Martin was Louis XIV, and as for his wife—she chose to dress as Mary, Queen of Scots, in a black velvet gown with a vintage high lace collar, a satin petticoat, and a stomacher so ornate it appeared to be made entirely of gemstones.

And as expected, the jewels on display were extraordinary. Cornelia Bradley-Martin wore her holdings of the French crown jewels, of course, repurposing the Empress Eugénie's bracelets as choker necklaces, and adding her own diamond belt and tiara to the outfit. (It was a time in New York City nightlife when the tiara was a definitive wardrobe item—hosts would request tiaras the way contemporary invitations denote black tie. The headpieces started at $150, or $4,123 today, in the Tiffany & Co. catalogue.) In the days leading up to the ball, newspapers reported on the frenzied preparations of the rich: "There is no estimating the value of the rare old jewels to be worn at the Bradley Martin Ball. All the jewelers who deal in antiques say that have been cleaned out of all they had on hand, and people still keep calling for old buckles, snuff boxes, lorgnettes, diamond and pearl studded girdles, rings, and, in fact, every conceivable decoration in gems. . . . All this, of course, is outside of the costly jewels held as heirlooms by the old families of New York. These have been taken from safety vaults and furbished up for the occasion in such quantities that the spectator will be puzzled to know where they all came from." Mrs. Herbert Pell stunned in a gold brocade gown, trimmed with ermine, with jewels running down the train, a jeweled stomacher, and more diamonds in her hair. A woman dressed as Empress Joséphine attached a purple cape to her simple white gown with large diamond clips on either shoulder. Called for comment by the *New York Times*, Tiffany & Co. addressed rumors that invitees would feel pressure to supplement their real gems with rhinestones: " 'It is ridiculous to suppose,' said one of the head gentlemen there, 'that the quality of people who have these rare and costly gems would ever think of attending such a historic function in sham ornaments.' "

With this phrase—"the quality of people"—the Tiffany & Co. representative had inadvertently hit the nail on the head. No matter what she told herself, or her husband, or her friends, Cornelia Bradley-Martin didn't pay $9,036.45 to the Waldorf

($248,000 now)—including $2,991.70 on champagne alone ($82,250)—just to stimulate the economy. She wanted to prove, once and for all, that she was the quality of person who could successfully host a landmark party; many of her guests, as they scurried from jeweler to antique seller, from hatmaker to hairdresser, in the weeks leading up to February 10, had something at stake as well. For years, New York society had been defined by the "Four Hundred": the go-to list kept by Ward McAllister, Mrs. Astor's social secretary, said to be limited by the capacity of the Astors' ballroom. Mrs. Bradley-Martin, a member of the Four Hundred herself, hoped that in throwing the ball, she would draw her own wider, more outrageous social circle, with herself and her husband at the helm—and those who hadn't made Mrs. Astor's original cut wanted in.

The Bradley-Martins paid a high price for their ambitions. Even though the ball had arguably succeeded on one front—it gave much-needed work to plenty of city artisans, retailers, and purveyors—the off-key (and self-congratulatory) tenor of the evening, in the context of larger social suffering, was too grating for almost anyone with a public platform to ignore. The Bradley-Martins had thought the event itself would put an end to the criticism, but the effect was quite the contrary. They became a joke, the butt of satire, and shorthand for moral outrage. Adding insult to injury, the public scrutiny drew attention to their finances and the city's politicians responded by doubling their property taxes.

Not too long after the Bradley-Martin ball, the Bradley-Martins left America—for good. True to form, they threw themselves a goodbye party before departing for England, where their daughter lived with her husband, the Earl of Craven. But it was only dinner for eighty-six, and couldn't possibly compare to the splendor of the ball at the Waldorf. As the couple had learned, there was magic in a costume party. For one night, those in attendance got to imagine themselves to be as sparkling and special as the diamonds they came dressed in. Sparkling, special, and rare—or so the American diamond devotees thought.

2

The Seekers

HOW ONE SICKLY ENGLISHMAN TOOK OVER THE
DIAMOND INDUSTRY

South Africa

When Cecil Rhodes arrived in South Africa in 1870, he was a sickly boy of seventeen, eager for a life outside of his reverend father's crowded home in Hertfordshire, England. One of twelve children, the pale, fair-haired Rhodes struggled to stand out, dreaming of the education at Oxford University that posh London boys took for granted. But the world had other plans for Rhodes. The teenager suffered setbacks from his congenitally weak lungs and heart, and his quixotic temperament didn't serve him well when it came time to distinguish himself as a budding academic. His grades were inconsistent, and his father couldn't afford four years at Oxford anyway. So Cecil Rhodes finished high school and saw his future blunted. His parents encouraged him to pursue a religious vocation, but he couldn't muster any real enthusiasm for the church.

Then, his older brother Herbert wrote. Would Cecil like to join him on the cotton farm in South Africa? The dry, subequatorial air might do wonders for his lungs. To a boy who lived with the expectation of dropping dead at any moment, and who had barely any reason to leave his hometown, the letter appealed to an underfed sense of adventure. And so, he departed England for a distant colony, with little expectation or ambition beyond a teenager's impatience to move out of his father's house.

The decision proved to be the turning point of his life. One rarely knows he is fulfilling his destiny until afterward, when those first steps, in hindsight, look nothing short of inevitable. Historians have worked to make sense of Rhodes's legacy, but whether they come down on the side of mastermind or megalomaniac, Rhodes, who found a sense of purpose in South Africa, got his eventual wish either way: that he wouldn't be forgotten.

<hr />

Rhodes landed in a country on the verge of transition. In the seventeenth century, South Africa's arid landscape was settled by the Dutch who stumbled onto shore, queasy and malnourished, after a brutal ocean journey with the Dutch East India Company. The intention was to set up a port town where future company ships could dock and find rest, supplies, and medical attention on their way farther east to India and China. Over the next few years Cape Town came to life, and while many travelers established residence along the coast, others migrated toward the interior in search of farmland. The Boers (the Dutch word for farmers) found their home on the hard-to-till soil, without regard for the indigenous population that already inhabited it. They called the native Khoi the Hottentots, from their word for a stutter, after mistaking the native tongue, which made use of clicks and other unfamiliar tones, for a collective speech impediment. The language barrier was the least of their troubles; the relationship between the Boers and the Khoi was not primed for a future of mutual understanding and respect. The religious Boers believed wholeheartedly in the keeping of slaves and so, over time, they managed either to drive out or subjugate a vast number of Khoi people.

Fast-forward almost two hundred years: Fifteen-year-old Erasmus Stephanus Jacobs, whose Boer family has known no other home than South Africa, is playing outside with his siblings. By this time, the country is officially a British state, though the resident Dutch, who now speak a local dialect called Afrikaans, are loath to give up cultural control. During the Napoleonic Wars, France invaded the Netherlands, and Great Britain responded by seizing Cape Town. However, like a child swiping a toy on principle and then abandoning it, the ruling country assessed South Africa—with its dirt-cracked land and impoverished residents—and quickly turned its attention elsewhere. The British left Dutch rule in place and focused the colonialist eye on other areas of sub-Saharan Africa that appeared more imminently desirable. It remained this way until 1806, when, once again prompted by French aggression, the British swept in and took on the country's entrenched racism by shutting down the Dutch East India Company's slave trade and introducing missionaries who counseled gentler treatment of the Khoi.

The Boers did not take kindly to these changes, and by 1867, many have scattered. Erasmus Jacobs and his family live just north of Hope Town near the bank of the Orange River. They keep rocks handy to play the traditional game Five Stones, similar to jacks, and one sunny spring day, an unusual white one has found its way into the collection. They begin to play, and a neighbor, Schalk van Niekerk, happens to notice the way the *mooi klip*—pretty pebble— catches the light. He asks the kids about it, and Erasmus explains that he found it one afternoon while taking a rest near the water, where he saw a glint on the ground; put the stone in his pocket; and, when he returned home, gave it to his little sister. Van Niekerk requests a closer look, and the Jacobs family watches, bemused, as he inspects it; he runs his fingers over the surface and then does something very strange—he drags it against a windowpane. He offers to buy it, but Mrs. Jacobs, remarking how silly it would be to charge a friend for a pebble and noting his surprising flash of energy, assures him it's his to keep.

As he walks away, Van Niekerk feels like singing. The stone scratched glass—from a book about unusual stones, he knows that diamonds are the hardest of all gems, and thus he's certain this is a diamond he has in his hands. But as the days pass, he second-guesses himself: Did he press too hard on the window? Won't any old thing scratch glass if you really dig it in? He decides to sell the stone and get whatever he can for it before someone worldlier shows up with evidence that it's not a diamond. In the end, it's an easy choice: he'll get more money for a maybe-diamond than he will for a local boy's river rock. Van Niekerk has a friend, a trader named Jack O'Reilly, who takes it off his hands for a few pounds. O'Reilly then tries to unload it in town, but he's met with universal scorn: diamonds aren't found in South Africa, everyone knows that. This is nothing more than an attractive skipping stone.

O'Reilly manages to persuade one person, who takes it to an

amateur mineralogist in Grahamstown, Dr. W. Guybon Atherstone. Atherstone proclaims Van Niekerk's *klip* a diamond, and the next day, the town is buzzing. After that, the diamond eventually finds its way to Garrard—the crown jewelers—in London, where it's confirmed to be the genuine article. Still, experts roll their eyes at the possibility of any more where it came from. In all of time, significant stores of diamonds have been found only in India and Brazil and this—the Cape Diamond, as Garrard christens it, later known as the Eureka Diamond—is likely no more than a gag, put on by someone with a financial stake in South Africa who planted the gem in order to attract widespread attention. Of this, they're enormously, scientifically certain. Given the quality of the earth in South Africa and how different it is from other places where diamonds are found, this stone cannot be a product of its environment—rather than having emerged organically from the land, it's more likely to have tumbled from the sky.

Another English mineralogist travels to South Africa and confirms Garrard's opinion. Even diamond fever can't withstand these harsh doses of reality, dispensed by the most qualified professionals, and eventually, it cools as South Africans come to accept that the Cape Diamond was a fluke. But then, another unlikely discovery changes the course of the country's history. In March 1869, a Griqua (biracial, usually the child of a Boer father and a Khoi mother) shepherd finds an enormous stone and brings it to none other than Schalk van Niekerk, who has gained a reputation as a man willing to pay a high price for diamonds. Sensing opportunity, the shepherd demands just that: five hundred sheep, ten oxen, and a horse. Van Niekerk agrees, and this time, he himself takes the rock to Cape Town, where he asks a jeweler, Gustav Lilienfeld, for an appraisal. Sure enough, this time there's no doubt that the hulking gem is exactly what it appears to be: an 83½-carat diamond. Suddenly, Van Niekerk's outlay of livestock doesn't seem quite so foolish. He sells the diamond to the jeweler for 11,200 pounds, and Lilienfeld christens it the Star of South Africa.

He puts it on display in Cape Town. Then, after the stone is cut and polished by the professionals in London, Lilienfeld sells it to an English nobleman, the Earl of Dudley, for more than double the price he paid. The sale attracts the attention of the international media, and with that, rumors of diamonds in South Africa spread well beyond London and Cape Town. Promises of unearthed riches appeal to all manner of men, but especially the prospectors in California and Australia, who had gone chasing gold but, in both cases, had recently witnessed the end of the rush. Fortune seekers from all over the world set their sights on South Africa. In an improbable twist, the remote fields that the English once dismissed as worthless and willingly relinquished to the Dutch, fields where the Boers live humbly in wood huts, sleep under animal skins, and cook over outdoor fires—those very same fields are, by 1870, widely regarded as the site of the world's next treasure hunt.

◇◇◇◇◇◇◇◇◇◇◇◇◇◇◇◇◇◇◇◇◇◇◇◇◇◇◇◇◇◇◇◇◇◇◇◇◇

When Cecil Rhodes made the trip to meet his brother, he was one of many Europeans who descended that year on South Africa. As young Rhodes adjusted to the rhythms of farm life, countless others disembarked at the Cape and then began the monthlong trek toward Hope Town and the Orange River, where Erasmus Jacobs's *mooi klip* had been found, and farther still, to the river's largest tributary, the Vaal. Their journey wasn't just long; it was difficult: foreign prospectors weren't prepared for the mercurial sub-Saharan climate, which frequently shifted from blisteringly hot to soaking wet with

its flash thunderstorms. Still, they pushed on, inspired by the spec-
ter of diamonds. Of course, very few of these men had ever seen an
uncut gemstone in real life, and when they arrived at the so-called
diamond fields, they were shocked to discover that the ground was
splashed with shiny pebbles of all kinds. Among chips of sparkly
sandstone and quartz, as well as rough agates, jasper, and garnets,
the diamonds—if they were there at all—were camouflaged. More,
the beds along the river were already swarmed. Boers, who had
come from all directions and pitched their tents, stood stooped by
the water, inspecting rocks. Local tribesmen—Khoi, Griquas, and
Koranas—had gathered as well, in the hope that the Boers, unac-
customed to labor, would pay them to dig and forage.

The immigrants joined the rush. They'd already traveled great
distances, and even though competition was stiffer than expected,
the playing field was relatively flat: there wasn't an experienced diamond miner among the bunch, and hardly anyone knew, beyond
the broadest terms, what a rough diamond actually looked like.
Until a digger had the chance to hold one in his hand, the diamond
was only a dream, an intangible; it was the thing that pushed men
to work after the sun got too hot, and forced them to crouch and lean
into the earth, even when their backs hurt. As Gardner Williams,
author of *The Diamonds Mines of South Africa*, put it: "Digging for diamonds never becomes dull drudgery, for there is always the glittering possibility in the mind's eye of upheaving a king's ransom
with the turn of a shovel, and it is far more exciting to a novice than
mining for gold or any other minerals." But the results of their
labors weren't always as fruitful as their fantasies. Quite often,
prospectors mistook quartz for diamonds. Unknowingly, they destroyed real diamonds by smashing them with hammers, thinking
the genuine stones would be indestructible.

Yet here and there, through a system of trial and error, a digger would find a diamond, and that success sent a jolt of energy across the shores as run-down men felt their hopes renewed. Their methods were unsophisticated, yet they were similar to strategies used by successful miners in historic diamond locales. The South African prospectors followed the Vaal River and dug, washed, and sorted in the manner of their predecessors in India and Brazil. In the hottest weather, they braved burning winds that carried debris from the diggings, stuffing up their noses and stinging and inflaming their eyes. They stayed despite modest discoveries: about 30,000 English pounds' worth of diamonds in 1870, which amounted to roughly 60 pounds per worker. (That's approximately 6,380 pounds per person in 2014, or $9,788.) Weak numbers, however, failed to deter the barrage of ambitious travelers who continued to trawl the riverbeds, and before the year was out, two diamond towns (Pneil and Klipdrift) and two diamond papers (the *Diamond News* and the *Diamond Field*) were founded. The predominantly male, multiracial community was a hotbed of aggression, and for some, the ever-increasing competition drove them away from the Vaal, toward the flat farms and occasional rocky hills that Afrikaners called the *kopjes*.

Those who wandered from the pack were rewarded, and during 1871, various new digging sites—remarkable because they were "dry," meaning unexpectedly distant from water, where diamonds, historically, had always been found—came to life. At a farm called Jagersfontein, owned by a widow, the overseer found a fifty-carat diamond, attracting the attention of the diamond flock. Glad for the chance to make a little extra money, the widow charged two pounds a month for the right to dig on her land in a twenty-square-foot patch. Other landowners followed suit, charging rent as well as a percentage of the diggers' findings. Thanks to the two diamond publications and the single-minded focus of South Africa's still-swelling population, word of any fresh successes traveled fast, and a farm that hosted just a handful of industrious prospectors soon was crowded and pockmarked with holes. That's what happened at a farm called Vooruitzigt, where a pioneer miner hit pay dirt. Men flocked to the site owned by a Mr. De Beer and negotiated with his son-in-law, who reluctantly sold off thirty-square-foot claims at the going rate of 25 percent of any diamonds removed from his family's property.

A few months into prospecting on Vooruitzigt, a Boer from Colesburg named Fleetwood Rawstone turned his attention to one of the kopjes that dotted the farm. Previous efforts by other miners to explore underneath its elegant camel thorn trees had proved un-satisfying, but in July of 1871, Rawstone and his friends had an in-kling that no one had tried hard enough. They sent a servant with a shovel to investigate, and he returned casually holding a two-carat stone. Immediately, Rawstone's crew staked their claim, and news of the productive Colesburg Kopje echoed across the land.

After that, Vooruitzigt was inundated with prospectors, and it became the site of the New Rush—as opposed to the first one on the banks of the Vaal—also known as the De Beers Rush, and the epi-center of the dry diggings. Located on British-controlled land, the central town that cropped up in the area between the major dia-mond farms was named Kimberley, after the British secretary for the colonies. Colesburg Kopje, which justified Rawstone's instinct by yielding thousands of high-quality, glassy stones, was rechris-tened the Kimberley mine. But even with a substantial financial stake in the diamond fields, the farm owners were often displeased with the unexpected turn their lives had taken, as if a biblical plague of brutish, dirty explorers had been sent to South Africa to uproot their quiet lives. Not only was their land crumbling underneath them, but the Boer temperament—sturdy, contemplative—didn't mix well with the kind of people who were inclined to leave everything behind in the pursuit of outrageous fortune. They were criminals, deserters, runaways; the town of Kimberley was a veritable Sodom and Gomorrah. Time away from rules, routines, and family made the men coarse, and gambling, sex with prostitutes, and drinking were the preferred pastimes. Johannes Nicolaas De Beer, who owned Vooruitzigt and resented his son-in-law's decision to rent to prospectors in the first place, made the only choice he felt he could: he sold his farm to a group of investors for 6,000 guineas. It was roughly half of what the Earl of Dudley had paid for the Star of South Africa, but for De Beer, money wasn't the object. The sale was an effort to protect his family from any further exposure to the blanketing darkness: the greed, sin, and violence. As he packed his family into a wagon, Johannes De Beer had no idea what his name would one day be attached to, or what legacy it would leave behind.

◇◇◇◇◇◇◇◇◇◇◇◇◇◇◇◇◇◇◇◇◇◇◇◇◇◇◇◇◇◇◇◇◇◇◇◇◇◇

The frenzy of the dry diggings reverberated all the way to Natal, where Herbert Rhodes's farm was located, more than seven hundred kilometers from Kimberley. The two brothers could not have been more different; Herbert was extroverted, animated, and impulsive, and when he heard about the riches coming out of the ground nearby, he thought little of taking off to explore the diamond fields and leaving his shy and inexperienced young brother in charge of his land. But unexpectedly, Cecil adapted well to his new role, and during his first season at the helm of the cotton farm, he markedly increased their profits. The money inspired him, and soon Rhodes had a plan for his time in South Africa: to live out a thwarted childhood wish and save enough to go back to England and bankroll an education at Oxford. In the meantime, he was enjoying his new lifestyle. His health had improved—though the possibility of a devastating relapse was never far from his mind—and he enjoyed the predictable rhythm of his days, with his sunlit hours spent overseeing the fields and his evenings passed in scholarship. The measure of power he had as the head of a small but thriving business was appealing, too.

He wrote to Herbert with good news about the farm, and his brother replied with colorful stories about the quest for diamonds. Cecil was spellbound. For a young man who likely had little knowledge of gemstones beyond the fantastical legends surrounding the Koh-i-Noor—the "Mountain of Light," a stone its owners were willing to die for—Herbert's letters sparkled with the romance of actual jewels. "You cannot understand what an awful enticement the diamonds are," he wrote home to his mother, the family member with whom he shared the easiest rapport. Without ever having been to Kimberley himself, he instinctively channeled the thrill that kept prospectors motivated: "Any day you may find a diamond that will astonish the world."

And so, in 1871 Cecil Rhodes followed his brother once again, this time to the New Rush. Herbert had purchased three claims on the De Beers' land, and Cecil—now over six feet tall, with intense, deep-set blue eyes and a foppish, light-colored wardrobe better suited to the great halls of learning than the dusty plains of South Africa—wanted in on the action. After all, diamonds would likely pave a faster route to Oxford than cotton. But if his lofty ideas about mind-boggling stones propelled him through the journey to Kimberley, the day-to-day drudgery of diamond mining quickly proved a reality check. Like most of the white prospectors who arrived in Kimberley with money in their pockets, he hired a group of native workers to do the digging for him. This meant that all day long he sat on an overturned bucket, waiting for his crew to emerge with a remarkable find. Other men, like Herbert, had an easy time making friends and talked, laughed, and played practical jokes to pass the time. But Cecil had no interest in socializing. He had little interest in women, either, so he was the odd man out in conversation even when he tried to be polite. As the other overseers chatted, he scowled and read original Greek and Latin texts, including his favorites like the *Meditations* by Marcus Aurelius, over and over. He stood out—Rhodes's delicate constitution, along with his brooding temperament, made him a very odd fit for the New Rush. Then the rainy season started, and even the work itself became more difficult.

The mines, which every day grew deeper, filled with water. By now, claim owners realized that diamond-bearing ground didn't stop just a few feet below the surface, as the original Vaal River prospectors had assumed; in fact, some of the richest deposits rested many layers beneath the earth, after the soil changed from red at the top, to yellow, to blue. This blue ground—eventually dubbed kimberlite—when exposed to the elements, crumbled like ancient remains and was rich with diamonds. This meant that some claims extended more than sixty feet below, with native workers climbing up and down ladders, filling and emptying buckets—labor that was dangerous enough already without the imminent threat of drowning. The rains paralyzed the mines, and the only way to resume work was to pump water out of the holes.

The machinery best suited to the job was hard to come by, and Cecil Rhodes, whose capitalist spirit had been awakened by his time in South Africa, had an idea; he rushed to approach the only pump owner in the vicinity and buy him out. He overpaid—the owner was understandably reluctant to sell—but made back his money quickly by renting the pump to desperate claimholders whose efforts had been halted by bad weather. Once one man drained his mine, the rest invariably followed, as the hunt for diamonds always felt as if it was one big discovery away from extinction. Rhodes made enormous profits, which he invested almost exclusively in the prolific De Beers mine. By this time, claims were quite expensive and individuals were taking buyouts from local and European companies for their shares. But Rhodes was cultivating his talent on two fronts: his head for strategy (especially when it came to besting opponents he considered less worthy than himself) and his knack for identifying his own weaknesses, which led him to seek out allies in business who had complementary strengths. As awkward and unfriendly as Rhodes could be, he was also smart, fiercely determined, and guided by a sharp inner compass that he trusted unfailingly. His unmistakable confidence—which bordered, at times, on full-blown narcissism—proved deeply charismatic to certain kinds of useful, loyal men.

Over the next few years, Cecil Rhodes became the second-largest claimholder in the De Beers mine, and made a name for himself as an important figure in Kimberley. Restless Herbert, on the other hand, left their diamond holdings to pursue rumors of South African gold deposits—only to burn to death when his tent caught fire one chilly night. Meanwhile, Cecil never lost sight of his original goal: to make enough money to spend four years at Oxford. Sure enough, he left England as a scrawny, introverted boy and returned a wealthy, confident man. He enrolled in Oriel College, leaving trusted colleagues in charge of his mining interests.

He didn't stay at university for as long as he would have liked. His initial stint lasted just a few months because his health deteriorated in Oxford; Herbert had been right that the South African climate was a comfort to his lungs. His money didn't last as long as he had planned either: part of it went toward tuition and living expenses, and much of what he earned was allocated to maintaining his position in Kimberley. But this didn't matter—eventually he figured out a way to split his time between England and South Africa, and his studies at Oxford were just as eye-opening as he ever could have hoped. Back in his mother country, surrounded by the best minds of his generation, Rhodes came to an important conclusion about his future and what he really wanted. With the accomplishment of attending a prestigious college under his belt, Rhodes understood that he needed a bigger, more sweeping ambition to motivate him. In the short time he had on this earth, he wanted to know that he'd made a difference. Suddenly, he understood that he required nothing less than a diamond fortune. And he knew exactly what he would do with the money once he earned it.

Rhodes was likely inspired by the critic John Ruskin, who was then lecturing at Oxford. Ruskin was a staunch British imperialist who argued that colonial aggression was justified if it meant restoring the nation to its former glory. Sure enough, Rhodes—an odd student who wore old dirty linen but frequently carried loose diamonds in his pocket—started to draft his own philosophical treatise, which he called "A Declaration of Faith." By this time it was 1877, and he had traveled back to South Africa and then returned to Oxford again, chipping away at his degree while simultaneously maintaining his business. He'd crystallized his views of the non-English Africans: "I contend that we are the finest race in the world and that the more of the world we inhabit the better it is for the human race. Just fancy those parts that are at present inhabited by the most despicable specimens of human beings what an alteration there would be if they were brought under Anglo-Saxon influence." Rhodes knew exactly how he intended to accomplish this: "Every acre added to our territory means in the future birth to some more of the English race who would otherwise not be brought into existence."

But Rhodes didn't simply want to inspire imperialist action, like John Ruskin; rather, he fancied himself the hailed conqueror, like Napoleon. In his mind, he was a self-appointed Alexander the Great, with no formal backing from the British crown. He planned to use diamonds as a way to finance a new era of English conquest and acquisition. Eventually, he believed, he would become the toast of London, the pride of his queen—and the ninety-eight separate claimholders still standing in Kimberley certainly weren't going to stand in his way.

On the contrary: they were the linchpin of his plan. As it stood, each mine was a patchwork of separate holdings, and Rhodes recognized that this model was unsustainable. First of all, with so many different interests competing to sell their diamonds, prices would inevitably plummet. Second, as each group dug deeper to get at the precious kimberlite, it was becoming more difficult to maintain the structural integrity of each claim and keep claims distinct. Rhodes set about implementing a strategy to consolidate the mines with exactly the kind of sharp, single-minded focus that had eluded him as a child. Unfortunately, someone else had a similar notion, and a face-off was imminent.

◇◇

While Rhodes was studying contemporary philosophy, Barney Barnato (originally named Barnett Isaacs) was enjoying everything the permissive town of Kimberley had to offer. He was as brash as Cecil Rhodes was diffident; he dressed in loud suits, had a background in acting and bartending in London, and had first made a living in Kimberley by passing off boxes of average-quality cigars as Havanas. But like Rhodes, he had followed his older brother to the diamond fields and started questioning the wisdom of the claimholder system. Also like Rhodes, he was ambitious and wanted to be the one left holding the cards when his competition folded. However, he didn't concern himself with lofty justifications like the reinvigoration of the once-glorious British Empire. He was solely fixated on the Kimberley mine—and on augmenting the contents of the ever slightly underwhelming Barnato purse.

Sure enough, Barnato earned plenty of money, having been one of the first men in the region to suspect it might be worth the work of cracking into that odd-colored blue ground. For Barnato—a Jew who grew up poor in England and who earned money doing everything from reciting Shakespearean soliloquies while walking on his hands to challenging much bigger fighters than himself in the boxing ring—the allure of success was pure and simple: he wanted to be taken seriously. He was a gifted showman with a survivor's instinct for salesmanship, but in South Africa as in England, he lacked the refinement of some of his peers. Quite the contrary, Barnato had a history as a hustler and a former street busker, and his reputation as a clown followed him. Even as he founded the lucrative Kimberley Central Company and won a position on the town council, he struggled to earn the respect he felt he deserved. His work as a councilman led to important public health reforms, but still no one would let him forget that his candidacy had been widely regarded as a joke and that he could credit his victory only to the trail of payoffs he'd left behind him. To the locals, the little man in a top hat with a waxed mustache would always be a *kopje-walloper*: a secondhand diamond peddler suspected of trading in stolen stones.

For a time, Rhodes and Barnato peacefully coexisted, although the two mines—De Beers and Kimberley—were spitting distance apart. But neither man's grand scheme benefited from complacency, and before long, each was eyeing the other's holdings. Barnato had the stronger position, for a handful of reasons—not least that the finest-quality stones consistently emerged from the Kimberley mine—but Rhodes, who had been elected to Cape Parliament, had better political standing, as well as the good sense to seek out powerful advisers who could connect him with influential people. Two friends brokered introductions to the notoriously discerning venture capitalist Nathan Rothschild, who agreed to a meeting in London. Rothschild concurred that the very future of the diamond industry depended on there being only one diamond king of South Africa. He helped devise a strategy to oust Barney Barnato and pledged his financial support.

By then, Rhodes had come to the important conclusion that the price of diamonds could be maintained only by carefully controlling the number of stones available on the market at any given moment, and that his own company should absorb the competition—with the industry's best interests in mind, of course. He approached Barnato to convince him. The pair met at Barnato's office, and Rhodes annoyed his associate by quoting lengthy passages in ancient languages. "When he talked Greek," Barnato later told a colleague, "I

knew he was dotty. He thinks he's Kimberley's bloody Messiah."
Rhodes had misjudged his competitor, failing to impress him and
instead pushing his buttons and inflaming his inferiority complex.
Barnato refused to make a deal. Then, as an ostensible gesture of
goodwill, Rhodes invited Barnato to view a collection of stones from
the De Beers mine, as if to prove that its yield was just as good as
Kimberley's. Barnato arrived to find an assortment of high-quality
diamonds laid out on a table, already graded and perfectly arranged
according to their value. Rhodes offered the whole lot to him at a
very good price, knowing the former *kopje-walloper* couldn't resist a
bargain. Sure enough, Barnato agreed to the terms. Without hesi-
tation, Rhodes indulged in a celebratory gesture, tipping the table-
top and pouring the entire display into a ready and waiting recep-
tacle. He explained to the room that he'd always wanted to see what
a bucket full of diamonds looked like. But Barnato, bested, knew
the truth. It would take weeks for the Kimberley Central team to put
their purchase back in order. Rhodes had therefore ensured that
Barnato would play by his rules, selling the gemstones slowly and
keeping the price of diamonds stable without flooding the market.

It was a sign of things to come. Rothschild had identified a weak
asset in the Kimberley mine's portfolio: a corporate stakeholder
known colloquially as the "French Company," which Rhodes might
use to establish a foothold in Barnato's mine behind his back. As
Rhodes made the trip to France to float his bid, Rothschild went
about raising the capital required to reinforce it. Both men were
successful, and Rhodes returned from Europe ready to celebrate—
until he heard the news that Barnato, already a low-level investor
in the French Company, wasn't willing to go down without a fight:
he was going to make a counteroffer. In turn, an irritated Rhodes
threatened to outbid him ad infinitum, therefore driving up the

price, but he suggested a tempting alternative: What if the two men came to an agreement?

Rhodes's proposed resolution was almost too good to be true. Would Barnato withdraw and let Rhodes buy the company if he promised to turn around and sell it back to Barnato for nothing more than the difference between their two high bids? All he asked for was 700,000 shares in the Kimberley Central Company, which seemed more than reasonable given that Rhodes was offering to save Barnato a serious amount of cash. Barnato accepted the terms, even with the knowledge that Rhodes was a slippery sort with some very powerful interests behind him. In the end, it all came down to perspective. Where Barnato saw two worthy competitors playing a variation on a friendly chess game, Rhodes saw a high-stakes showdown in which cornering the opponent was the only acceptable outcome. After all, Barney Barnato was just one man standing in the way of Cecil Rhodes's sweeping imperialist vision. Once Rhodes leveraged the French Company into partial ownership of Barnato's, he instructed his investors to buy up shares in Kimberley Central, regardless of cost, until he had a majority interest. By the time Barnato realized what was happening, it was too late. He put up a decent fight but proved he wasn't as cutthroat as Rhodes, and eventually, the pair struck another bargain. They would liquidate Kimberley Central. Immediately, De Beers would purchase the company stock. In March 1888, Cecil Rhodes founded De Beers Consolidated Mines Limited, which reflected all of his acquisitions. He was officially the king on a diamond-studded throne. As for Barney Barnato: by relinquishing the crown, he had still found his way to fortune. His shares in De Beers were significant and hugely profitable.

After that, there was no stopping Rhodes. He bought up mine

after South African mine, putting him in control of no less than 90 percent of the world's diamonds. When he laid off workers in the fields and cut wages for everyone else, leaving many impoverished and without recourse, there wasn't even a political party to appeal to, as Rhodes had recently been elected South Africa's prime minister. He earned the nickname "Colossus," and under his rule De Beers and its associates marched farther north into Africa, subduing local tribes in the quest for gold mines that would round out their stores of treasure. He conquered a region called Matabeleland (now Zimbabwe) and called it Rhodesia, enjoying the double triumph of annexing new territory for his queen while seeing it bear his name. Although he was celebrated both at home in South Africa and abroad in England for his accomplishments, Rhodes was becoming a villain to native Africans, who lost not only their land but also their access to the resources that made it desirable. About this, the Colossus had no second thoughts; he viewed the indigenous population as "just children" and "emerging from barbarism." His was a classic Victorian, ethnocentric view; the journey took place in abject darkness, and he imagined himself, by way of English culture, as nothing less than the one holding the lamp.

This perspective might have been defended by highbrow philosophy, but it was also particularly self-serving for a man who stood to profit directly by subjugating his workforce. Since the early days in Kimberley, it had been clear that dark-skinned laborers were second-class citizens. The early claimholders instituted policies, like mandatory searches and curfews, that applied to native blacks—but not whites. In a supposed effort to curb illicit diamond buying, or "I.D.B.," Cecil Rhodes, at the helm of De Beers, tightened these rules so that native workers at the end of their contracts could be detained in barracks, where they were strip-searched, kept nude, and subjected to humiliations like having their excrement examined to make sure no one tried to smuggle a diamond by swallowing it. As time went on, it was considered a given that black workers were paid less and treated worse. But these inequities became politically entrenched in 1894 when Rhodes, as prime minister, approved the Glen Grey Act, which applied to newly annexed land where Dutch famers and native Africans were competing for space. The legislation changed the definition of land ownership so that it no longer extended to communal arrangements—which were typical among tribal groups—and honored only the more European model of one household per plot. At Rhodes's whim, countless black Africans lost their homes and faced one of two unpleasant options: they could either choose to relocate or become laborers for their new white landlords. For the so-called privilege of staying put—a "gentle stimulus," Rhodes said—the act suggested they could be charged a tariff. These laws were put in place and adopted elsewhere in South Africa, and ultimately provided an original blueprint for the injustices of apartheid.

Rhodes never gave up his quest for English dominance, and when he died of heart failure in 1902 at the age of forty-eight, the unmarried, childless tycoon left Oxford enough money to institute a substantial scholarship intended to bring the best and brightest from all over the world to the motherland. Despite his ruthlessness, Rhodes was the subject of many warm tributes, including one from the popular writer and unapologetic imperialist Rudyard Kipling, who penned a new poem for the funeral: "When that great Kings return to clay / Or Emperors in their pride / Grief of a day shall fill a day."

Further, Rhodes's contributions to the diamond industry would have widespread implications that continue to shape the market-

place today. Following his theory that the price of diamonds was
stable only when the number of rough and polished stones for sale
was very carefully controlled, Rhodes made sure that De Beers
monitored its output from the moment the diamonds came out of
the ground. The layoffs at the fields were the result of a significant
and deliberate decrease in production, and afterward, he started
stockpiling any excess output so that the illusion of diamonds
being rare—the key, he understood, to their value—was maintained. Rhodes recognized that his best customers, the American
rich, were enamored of the crystal clear gemstones and were willing to pay high prices for them largely because they believed money
and status gave them privileged access to the earth's most precious
treasure. In order to maintain this illusion, and to guarantee that
no one else challenged it, Rhodes gave De Beers a clear, if difficult,
goal: complete and total monopoly of the market.

His tactics, while draconian, worked. The price of diamonds
rose steadily, and from 1898 to 1902, American imports of gemstones more than doubled.

3

The Lovers

HOW YOUNG WOMEN CAME TO EXPECT DIAMOND
RINGS (PART I)

United States and Europe

I n 1902, 25,412,775.72 dollars' worth of diamonds and precious stones came through the Port of New York, with more than seven million dollars' worth still uncut. After the panic of 1893 subsided, the economy started to rebound, and the years following the Bradley-Martin ball saw a rise—with little thanks to the party at the Waldorf, unfortunately—in nationwide prosperity. Owing to America's increased spending and the upper class's preoccupation with jewels, the country became the world's most voracious consumer of diamonds, and stateside industry flourished as more businesses emerged to support that appetite. Diamond cutting, long an enigmatic art form confined to a skillful few who resided mostly in the cities of Amsterdam (Netherlands) and Antwerp (Belgium), found additional centers in Philadelphia and Manhattan. Aware of a distinct and sudden eagerness among the public to own precious gems, jewelers seized the opportunity to create even more reasons to buy, introducing Americans to the tradition of giving and wearing natal stones, also known as birthstones.

An opal for October, an emerald for May. Each month corresponded to a different jewel, and women were encouraged to wear their proper ones—whether in the form of earrings, a ring, or a brooch—with pride. Via the Blue Book, Tiffany & Co. popularized a historical list of natal stones, which its esteemed gemologist George Frederick Kunz traced back to Poland in the eighteenth century. The list was then standardized across the industry—after the implementation of changes that Kunz strongly disapproved of—by the National Association of Jewellers in 1912. Among those changes: December's ruby was swapped for July's turquoise, with the explanation that the cool blue stone better suited winter's chilly temperatures and icy palettes. The chartreuse peridot was demoted from September's stone to an alternative for August's smooth sardonyx. As for the diamond, the modern king of all gems and the designated birthstone of lucky April babies, Kunz felt its inclusion was altogether unsuitable: "As the diamond does not appear to have been known to the ancients and is not given in any of the lists of birth-stones before the last century, and as diamonds like gold and platinum, may easily be used as accessories to other stones, would it not perhaps be better to omit the diamond from the list?"

His peers, who counted diamonds among the most costly stones, and temperate April as a good month for shopping, clearly disagreed.

Jewelers offered the list of birthstones as a helpful guide for gift occasions like Christmas, Easter, and graduation, all promoted by the industry as the perfect moment to surprise someone special with a gem. But birthstones had another purpose, too, one that was loftier and more memorable than any present tucked under the Christmas tree could be. In recent years, a trend had cropped up with such persistence that young people could hardly be convinced it wasn't a vestige of time immemorial. In fact, anyone who wasn't doing it was regarded as pitifully behind the curve. It went like this: A young man asked a young woman to marry him. With joy in her heart, she wanted to shout her acceptance from the rooftops, but she waited. And waited. Finally, the would-be groom produced a ring.

The practice of giving an engagement ring had some historical precedent, but never before had its implementation been so widespread. Convention dictated that the ring include a gemstone of some kind—precious, ideally, but semiprecious in a pinch—and birthstones were considered a perfectly acceptable choice for the young man shopping alone who felt intimidated by the array of baubles at the jewelry store. A birthstone ring showed forethought and consideration—unless the young woman's birthday fell in October, because opals still carried suggestions of misfortune to the superstitious. More extravagant couples might opt for the Victorian DEAREST ring: a band with a rainbow of tiny stones that "spelled" the word "dearest" (diamond, emerald, amethyst, ruby, emerald, sapphire, topaz) from left to right. A single pearl was fashionable, too.

The engagement ring had become so popular that in 1894, the *Boston Saturday Evening Gazette* noted, "[A] crusade is being started against the engagement ring. One of the reasons given for its proposed abolition is that many girls become engaged for no other purpose than to add another ring to their collection and break off the contract as soon as it becomes convenient." Whether or not this was entirely true, ten years later in 1904, the *New York Times* published a "Trade Secret" implying that, crusade notwithstanding, little had changed. Says the customer: "Are those five or six wedding rings all you have in stock? Why you've got a whole tray full of engagement rings."

The jeweler replies: "Yes, Sir, and it will take that whole tray full of engagement rings to work off those five or six wedding rings."

What had changed in ten years was that, slowly but surely, one style of engagement ring had eclipsed the others as the all but universal standard. It turned out that among the birthstones, and within the DEAREST ring, one stone shone brighter than all the rest.

◇◇◇◇◇◇◇◇◇◇◇◇◇◇◇◇◇◇◇◇◇◇◇◇◇◇◇◇◇◇◇◇◇◇

When Franklin Delano Roosevelt—a fifth cousin of President Theodore Roosevelt—proposed to the president's niece (and his own distant cousin) Eleanor in 1904, he offered her a 3.4-carat diamond ring from Tiffany & Co. They'd met before at family gatherings, but the romance started on a train, when young Franklin, then studying at Harvard, spotted a more sophisticated, confident Eleanor than the awkward girl he remembered seeing more than three years before. She'd been attending school in London, and over the next four years the pair got reacquainted via long, intimate letters and the occasional in-person meeting. Eleanor was intelligent, articulate, and compassionate, and by the time she was nineteen and Franklin was twenty-one, they were ready to walk down the aisle. Franklin's controlling mother, however, opposed the marriage, and it took him almost a year to convince her that at his young age, he had already found the woman he wanted to be his wife. On October 11, Eleanor's twentieth birthday, the engagement became official when Franklin gave her a symbolic present: a diamond.

The ring's sizable center stone was cushion-shaped—square with rounded edges—and held in place by eight platinum prongs. The band was yellow gold, and decorating it on either side of the prominent gem were three smaller ones that lay almost flush with the finger. The ring wasn't just beautiful; it represented a fresh new style. In 1886, amply furnished with stones from the plentiful South African mines, Tiffany & Co. debuted the Tiffany Setting: a ring setting that used prongs to raise a single diamond off the hand. Whereas gemstones had previously been set into rings by way of a bezel setting—a metal collar that wraps around the jewel so that, traditionally, only its face, or "crown," is exposed—prongs showed off more of the diamond and allowed it to sit higher, where it attracted light. The setting proved to be a huge success and a signature item for Tiffany.[*]

Charles Tiffany picked up on the engagement ring trend and must have understood that the diamond, suddenly available in unanticipated quantities, would surely surge in popularity along with any new cultural emphasis on jewelry. And indeed it did; in November 1904, just a month after the Roosevelts solidified their engagement, the *Chicago Tribune* reported that "the engagement ring is getting to be so indispensible to engagements in this day and age that the girl who admits her betrothal without at the same time shyly exposing the diamond solitaire that twinkles on her little left

· · · · · · · · · · · ·

[*]It was actually invented by Ferdinand J. Herpers, a Newark, New Jersey, jeweler who patented his "Improvement in Diamond Setting" in 1872. At the time, Tiffany & Co. relied on Newark factories for a good deal of its metal, including its award-winning silver, and even moved manufacturing operations there in the 1890s.

hand is extremely rare and hard to find. 'No diamond, no engagement,' may be set down as the modern watchword in matters of the
heart. Cupid alone is not sufficiently strong to ensnare and hold
the hearts of maidens fair to one man. He must have an assistant,
and that in the shape of a narrow band of gold set with a single diamond." Etiquette experts confirmed that the proper way to propose
was to come equipped with a diamond ring.

The spread of materialism, confined just a few decades prior
to the upper classes, was remarkable, and even more so given that
before the late nineteenth century, the number of examples through
history of a man proposing with a diamond ring could be counted
on one hand. Thus young American ladies in the early twentieth
century seemed to earn the disapproval expressed just a few years
before by Lord Randolph Churchill (the former British chancellor
of the exchequer, and Winston's father) when he visited Kimberley for the first time in 1891 and bemoaned that the pitted mines
were dug "all for the vanity of women." Yet it wasn't always this way;
women hardly ever wore diamonds at all until the fifteenth century.
Before that, they were the charms and ornaments of men. In South
Asia, Mughal kings and princes valued them for size rather than
sparkle, and wore them uncut—silvery but smooth like sea glass—to
ward off evil spirits. In Europe, men wore diamonds on the battle-
field as amulets that they believed would keep them out of harm's
way. Men were willing to murder one another for possession of the
right jewel. The Koh-i-Noor, for instance, reached a high level of
public infamy as much for its bloody past as for its size. It was once
said that whoever owned the Koh-i-Noor ruled the world, and its
owners went about defending it with the same fierceness they'd
muster for the throne to the kingdom of earth. Aga Mohammed
Khan, who ruled Persia in the early nineteenth century, tortured

his rival mercilessly to determine its whereabouts—all for nothing, because the victim didn't actually know where it was. It turns out the diamond was not in Persia at all but in Afghanistan, where rising Sikh leader Ranjit Singh starved its owner—a man who himself had imprisoned his brother on his route to power—until, half crazed with thirst and hunger, he relinquished it to Singh to save his own life.

The first woman said to wear diamonds was locked in another kind of competition. The petite Agnès Sorel, blessed with large eyes and a Cupid's bow mouth, was the mistress of King Charles VII of France. He'd been crowned in 1429 after Joan of Arc followed a religious vision to jettison the British at the Siege of Orléans; by the time King Charles spotted Agnès Sorel and started pursuing her, Joan of Arc had been burned at the stake. Sorel played coy and used her influence to persuade the king to oust the British, who since the Siege of Orléans had recovered their position in Normandy, once and for all. It was a battle fought for love (or something like it), and when Charles rode victorious into Paris, Sorel followed right behind him, with only one woman given a more prominent position in the entire entourage: his wife.

It wasn't long, however, before the court acknowledged Sorel's true standing. Marie of Anjou was the queen, but Agnès Sorel was the king's amour. Soon, she was also the mother of his children; she gave birth to three daughters, who played among the queen's fourteen little princes and princesses. Her special place within the king's heart earned her other accoutrements of status. She was born noble, but to a family that had fallen out of fortune. As the king's mistress, she had access to the finery that had eluded her as a girl, and she dressed in beautiful silks and furs. When she started wearing diamonds, some members of the court found her too brazen, as if she had basked so long in the rays of the king's affection that she had come to believe she could stare into the sun. Others, like Jacques Coeur, the king's financial adviser and Sorel's friend, saw opportunity.

Coeur recognized that the pretty, twentysomething Sorel had the air of a muse about her, and used her as a fashion plate. A wealthy merchant, Coeur had ships sailing all over the world on the hunt for luxury goods to import into France like ermine, spices, exotic dyes, and gemstones. Through his work, he met Louis de Berquem (or Berquen, also known as Lodewyk van Berken), a diamond cutter living in Belgium who was experimenting with new, beautiful proportions for gemstone faceting. Coeur suspected Berquem's delicate work might appeal to the women of the court, and he relied on Agnès Sorel, who wore clothes well, to set a glamorous example.

The relationship between Coeur and Berquem was a two-way street; Coeur stood to profit from the artisan's work, while for Berquem, who was Jewish, the implied acceptance of a high-ranking member of Charles VII's court provided an important level of protection at a time when religious persecution against non-Christians was rampant. Jews got their start working with diamonds precisely because systemic prejudice in Europe prevented them from entering other, more desirable trades. Owing to old superstitions about the dangers of cutting diamonds, including a persistent myth that diamond dust was a deathly poison, few tradesmen were willing to risk modifying the rough stone. Jews living in the Middle Ages saw an opening. During the mid-fifteenth century, when Louis de Berquem was trading with Coeur, the art of diamond fashioning—the process of cleaving, shaping, faceting, and polishing a gemstone—was in transition. The earliest cutters concerned themselves mostly with symmetry, but by the time a young, eager Berquem traveled to Paris to study diamonds, the tools had improved, as had the craft. The best cutters were masters who could assess a rough stone and know exactly where to cleave it to minimize imperfections and maximize beauty. They used grinding wheels, slicked with oil and diamond dust, to cut a series of angled planes, or facets, into the surface in an effort to bring out the shine and take advantage of the play of light.

Berquem was especially interested in a new cut that would eventually be called the brilliant: a mathematically proportioned diamond intended to maximize reflectivity, or, in trade parlance, brilliance (the way white light bounces off the stone) and fire (the dispersion of rainbow colors). His gemstones were more prismatic than any that came before him. When Jacques Coeur saw his work, he must have immediately understood: How could Agnès Sorel resist a jewel that would allow her to radiate rainbow colors, as if she'd channeled the heat of the universe? Once they'd glimpsed her power, and seen a king dangled around her pinkie finger, what woman of the court wouldn't want to look just like her?

Coeur's instinct was a good one. And all three—Berquem, Sorel, and Coeur himself—were immortalized eventually. Agnès Sorel posed topless for Jean Fouquet's *Virgin and Child*, before she died mysteriously of mercury poisoning at the age of twenty-eight. After he was "discovered," Louis de Berquem was hired by Charles the Bold, the Duke of Burgundy, to cut some very large stones. Two of these remain quite famous: the Florentine, which was said to be unrivaled in beauty but which disappeared after World War II; and the Beau Sancy, which was eventually given to the pope. Berquem, who trained countless young faceters, and whose work led to the establishment of a guild of diamond cutters in Belgium, leaves his legacy behind by way of a statue still standing in Antwerp; he's depicted holding a diamond. Jacques Coeur got a statue, too, in his hometown of Bourges, France, but not before he was accused of poisoning Agnès Sorel, imprisoned, and stripped of his riches. Eventually, his name was cleared, but he died soon afterward.

Without this trio, would the diamond engagement ring have found a foothold among English and American youth more than four hundred years later? At the very least, it was thanks to Sorel and Coeur that fashionable women started wearing diamonds. As for Louis de Berquem, his contribution to the brilliant cut inspired future fashioners who continued to experiment with diamonds and light. The cut found its perfect home in the Tiffany Setting, which was designed to feature a very specific stone: one with a conical pavilion (bottom point), a precise ratio between the girdle (the stone's widest measurement) and the table (the flat plane on top), and a carefully determined number of facets. The brilliant is carefully shaped to catch light, and the Tiffany Setting, with its supportive prongs, gives it a boost by raising it up a little higher. In other words: without the brilliant, there might be no solitaire.

∞∞∞∞∞∞∞∞∞∞∞∞∞∞∞∞∞∞∞∞∞∞∞∞∞∞∞

In 1477, not long after Agnès Sorel started wearing diamonds, Emperor Maximilian proposed to Charles the Bold's daughter, Mary of Burgundy, with a diamond ring, perhaps in part to garner admiration from her gemstone-loving father. Prior to their engagement, the practice of giving some kind of betrothal ring—diamond or anything else—had come into and out of fashion. It dates all the way back to the second century AD, when Romans offered their brides-to-be rings as a form of matrimonial insurance; before a group of witnesses, the man gave his fiancée this tiny yet valuable gift as a guarantee that they would eventually wed. Later in time, the betrothal ring disappeared in favor of a wedding ring, presented at the altar but still traditionally worn only by the wife. These rings were typically simple, made of silver, gold, or iron, and like the engagement ring served two distinct functions: to symbolize an exchange of property and to mark the woman as "taken." (The so-called double-ring ceremony, in which the wedding vows accommodated rings for both parties, came much later and wasn't universalized until the twentieth century, when in London, women vociferously complained that they couldn't tell a single man from a married one just by looking at him—and that this led to all sorts of unpleasantness.) The Puritans exchanged practical thimbles along with their plans to marry.

In other words: betrothal and wedding rings, no matter how beautiful or carefully selected, have always been indicative of the transactional nature of marriage. The husband pledges fidelity and financial support; the wife, in turn, promises that, physically and emotionally, she'll remain true. With that said, suitors have for a long time managed to make room for romance amid those other, more businesslike implications. Many cultures identify the third finger as the proper place for a lover's ring; this convention is inspired by the ancient Egyptians, who believed a vein started in the tip of that finger and trailed a direct blood pathway to the heart. Victorians were partial to DEAREST (or, alternatively, REGARDS) rings, and in the nineteenth century, sentimental gimmal rings became popular. These were two conjoined circles often meeting in a heart or a set of clasped hands. At the engagement, the circles were separated so that man and woman could each wear one. The rings were then reunited at the wedding ceremony, where the wife took full custody of the set.

When the diamond engagement ring trend caught on at the turn of the twentieth century, it represented the perfect balance of romance and financial contract. Women at the time were encouraged to covet diamonds, so the act of a man buying one for the lady he loved was already inherently sweet. At the same time, diamonds weren't cheap, so they also made for good wedding collateral. In theory, a man who gave a woman a diamond ring would think twice about rescinding his offer of marriage. And that was precisely the point: a proposal wasn't a trifle. When, in the thirteenth century, an English bishop noticed a trend of young men courting young women with "rush rings," or rings made of straw, as a goofy getting-to-know-you prank, he pronounced that they would be regarded by the church as binding; this promptly did away with the fad.

Diamonds, of course, were considerably more expensive than straw. For a man in 1894, a single Tiffany-set diamond from Tiffany & Co. would run him at least $10 (the equivalent of roughly $260 today). Just ten years later, in 1908, the price had doubled to a baseline of $20 ($540 today) for the least expensive version. This represented a considerable outlay for the average buyer, who might eye the tiny $10 opal—unlucky or not—as a tempting alternative. But what about the hard-up guy who could barely afford to shine his shoes but whose girl had her heart set on a brand-new sparkling diamond? For him, there were options—but he had to look beyond Tiffany & Co.

He might put the Blue Book aside and reach for another catalogue, one that his mother would consult to buy groceries or that his father could turn to if he wanted an affordable new shotgun. In June 1906, the Chicago department store Sears, Roebuck and Co. became a $40 million—equivalent to over $1 billion in 2014—player in the markets after Goldman Sachs & Co. and Lehman & Co. agreed to finance its transition to a joint-stock company. The news rumbled through Wall Street as if a giant had tiptoed up and then crashed through the front door. Just twenty years before, the business's cofounder and president, Richard Warren Sears, had been a humble telegraph operator on the Northern Pacific Railway, a transcontinental line that ran all the way from Minnesota, his home state, to the West Coast. But Sears had grander aspirations. In his time working for the railway, he'd gotten to know all manner of people, and had studied their needs, buying habits, and inclinations in the hope that one day he'd start a business that appealed to the farmers, railroad workers, and hardworking folks he regularly encountered. He'd also become interested in the mail-order format, observing that, between work, family, and chores, many of his peers had no time to shop.

The 1894 Sears, Roebuck & Co. catalogue had touted "The Cheapest Supply House on Earth." Essentials carried the store through the lean times, and by the end of the decade, it was able to close sales on little luxuries like china and silverware again. The Spring 1899 "Big Book"—so called for its size and massive page count—showed pages of diamond rings for the bargain-seeking suitor to choose from. They started at just $4.50 (approximately $122 in today's dollars) and ranged up to $121.50 ($3,326), though the store offered a good selection in the under-$10 ($275) range. The stones were genuine, imported from Amsterdam. Just like the rest of the merchandise—a $7.95 mandolin, a $7.85 baby carriage—a diamond ring could be ordered by mail, and then soon it was delivered to the post office, where an eager young man could inspect it and decide whether or not he was ready to make the purchase. There were no snooty salesmen involved; there was no personal pressure to buy. Sears, Roebuck & Co. understood that what it lost, in comparison with a house like Tiffany & Co., it gained in everyman appeal: "We are slaughtering the price of diamonds, as we have slaughtered the price of watches," it boasted in the Fall/Winter 1896 catalogue. "Don't pay the retail price for diamonds. At our prices, the goods are always as good as the money. They never wear out, never change, always the same, but heretofore they have been sold at such unreasonable prices that few people could afford to wear them."

Customers were seduced by this straightforward language and assertive sales pitch, and for a time, Sears, Roebuck & Co. could claim to be one of the biggest diamond sellers in the country. But it wasn't just R. W. Sears who aggressively courted business away from proprietary jewelers. In the late 1890s John Wanamaker, founder of the groundbreaking Wanamaker's department stores in New York City and Philadelphia, ruffled feathers when he expanded his jewelry departments and made a fevered play for one-stop shoppers with budgetary concerns in mind. "The days for selling jewelry at double its cost had ended," the very first advertisement for the new department proclaimed. How could Wanamaker afford to cut prices? Leveling with the public in the ad, he went on to explain that jewelers charged a premium for brand-name items; more, they were forced to overcharge to make up for the fact that the peak jewelry-buying season was extraordinarily brief, pegged almost entirely to the winter holidays. Department stores, as opposed to jewelers, he reasoned, could afford to spread the burden of sales across a variety of year-round merchandise. (In a rancorous response to Wanamaker's ad, published in the jewelry trade magazine *The Keystone*, one author quipped: "Without commenting on the foolishness of this statement, we will say that John Wanamaker has secured, at advanced salaries, some very excellent talent for its jewelry store, and we doubt whether their prospective removal to the necktie or glove counter in January will be looked upon by them in the light of promotion.)

Jewelry professionals felt threatened, and rightfully so. Even the *Keystone* author had to admit that Wanamaker's unfussy approach to selling jewels had its moments of wisdom. In the same ad, Wanamaker invited customers to feel free to browse in the department without any intention of buying. In other words, those who might have felt curious about jewels but weren't prepared for the intense one-on-one service of the local jeweler, or who were intimidated by the luster of Tiffany's, now had a welcoming place to go. The critic tipped his hand: "The man or woman who is given an opportunity frequently to examine jewelry stock will, sooner or later, means permitting, succumb to the passion for possession." Ironically, in an intensely competitive climate characterized by a greater number of retailers than ever before, everyone—from Charles Tiffany to Richard Sears to John Wanamaker—seemed confident that the merchandise would ultimately sell itself.

With diamonds being sold in department stores, and a thriving new tradition that had every young bride waving a twinkling ring on her left hand, there could be little doubt that diamonds—once the exclusive property of the royals and the rich—were being democratized. Everyone, for a spell, could have a tiny taste of status—which, as anyone with a real claim to it knows, means death by a thousand nibbles, a status symbol's downfall. Cecil Rhodes understood that a gemstone shines brightest when reflected in the eyes of someone who can't have it. So what was a socialite to do when, suddenly, her Tiffany diamond ring didn't look all that different from the one brandished by a farmer's daughter? The answer was simple, time-honored, and true.

Go bigger, of course.

Chapter 4

The Show-Offs

HOW A SINGLE STONE BEWITCHED ITS OWNERS—
AND CAPTIVATED AMERICA

Washington, DC, and Paris

Evalyn Walsh McLean never saw a diamond she didn't like. The daughter of a Colorado gold miner who struck it rich and moved his family to the moneyed social circle of Washington, DC, Evalyn Walsh grew up admiring the well-dressed socialites and politicians' wives. There was Mrs. Potter Palmer, a family friend and the young wife of a millionaire merchant more than twice her age; her beauty and social graces elevated the couple within society. Evalyn always remembered the feeling of holding Bertha Palmer's hand as a young girl, the rough edges of Palmer's diamond rings rubbing against her fingers and sending an electric shock to her stomach. Evalyn was a tomboy who nevertheless aspired to be an actress, to dress in silk and lace and to wear fine jewelry just like Mrs. Palmer. She found herself easily influenced by those women she considered to be most glamorous. A few years later, when the film star Edna May wore her wavy hair parted down the middle and covering her ears, Evalyn copied her, much to the annoyance of her parents and teachers. She refused to change it until her father bribed her with a diamond ring to push her hair back and style it neatly like the other respectable schoolgirls.

Thomas Walsh never could have known how a little diamond ring would change the course of his daughter's life. To Evalyn, a young, impressionable teenager, the feeling of owning her first real gemstone—admiring it on her hand, flaunting it in front of her friends—was just as satisfying and validating as she had imagined. More, the ring had come easy—she barely had to earn it. Her father spoiled his wife and two young children, sending mother and daughter on fanciful shopping trips everywhere from Wanamaker's to Tiffany's. Most important, the diamond attracted the attention of a boy: another rich Washingtonian, who spotted the ring and assumed it signaled an engagement. Edward "Ned" McLean was the pampered son of John Roll McLean, owner and publisher of the *Washington Post*. There was nothing in the world Ned wanted that his mother wouldn't eventually give him, and even the Walshes, in their sixty-room Massachusetts Avenue mansion, considered him overly indulged. When Ned saw Evalyn's ring, he proposed immediately, lest someone else claim the girl he had his eye on. Evalyn refused, of course—at fourteen, she was too young to marry anyone—but Ned stuck stubbornly in her mind like the sparkle of Mrs. Potter's jewels. No matter how many eligible young men she met or how many European royals courted her, Evalyn always came back to Ned, despite her mother's protests. In July 1908, when

Evalyn Walsh and Ned McLean were both twenty-two, they were married. Their fathers each pledged $100,000 (or $2.48 million today) for a wedding present, and with cash in hand, the pair set off on their honeymoon.

They traveled to Europe by ship, and their first move was to buy a Mercedes even though they had a shiny yellow Packard roadster to drive. Once they arrived in Amsterdam, they hopped into their new car and took off for Berlin, with their attendants, including a paid chauffeur, trailing in the Packard with their luggage. The young couple shopped their way through Germany, concluding the spree with the purchase of another Mercedes in Leipzig, seduced, perhaps, by the opportunity to buy a Grand Prix–winning Benz in its native country. They were flush, unencumbered, entitled—and besides, before she left Washington, Evalyn's father had encouraged her to pick out a special wedding gift overseas, a suggestion she was more than happy to take to heart. Still, the joyride couldn't go on forever. After visiting Vienna, Constantinople, and Paris, Evalyn and Ned were out of money and couldn't afford their hotel bill. They had to ask Evalyn's father to open a fresh line of credit. But even that didn't stop her from wandering into Cartier on the Rue de la Paix. Over the course of her trip, she'd bought a chinchilla coat, a gold traveling case, and two luxury cars, but the newly minted Mrs. McLean was still hunting down the perfect item to commemorate her marriage. The pair had already burned through over $200,000. At that point, what would buying another tiny bauble matter?

Or at least Evalyn felt it was her duty as an American socialite to look. By 1908, Cartier had established a reputation as the destination jeweler for the wives and daughters of the rich. In 1899, Alfred Cartier had moved the half-century-old business from a more modest shop on the Boulevard des Italiens to the fashionable Rue de la Paix, an airy cobblestone street that pointed toward the stately bronze obelisk marking the Place Vendôme. Alfred had taken over from his father, and along with his own three sons, Louis, Pierre, and Jacques, was in the process of transitioning the company from its original incarnation as a seller of little luxuries—jewels and other *objets*—to a respected jewelry manufacturer. The Rue de la Paix was the ideal address for the house that King Edward VII of England had recently dubbed "King of Jewelers and jeweler of kings." Since then, Cartier had been collecting royal warrants of appointment— designating preferred suppliers of various royal families. The new location appropriately reflected Cartier's skyrocketing reputation. Surrounded by retailers like Worth (a couturier), Guerlain (a perfumier), Vever (the jeweler), and Mappin & Webb (English silver-smiths), the French jeweler was in good company, situated among the kinds of businesses that attracted high-end customers. It was a place to wander and leave feeling lighter—particularly in the wallet. Wealthy young women traveled to Paris, where they ordered their gowns for the upcoming season from Worth and then, right away, dipped into Cartier and helped themselves to a fresh set of jewels to match.

The store was on the ground floor of a beaux arts mansion, ornately decorated with chandeliers and elaborate glass cases to complement the product. When Evalyn Walsh McLean walked in, the man working the floor took her in—a handsome, impeccably coiffed young woman—and immediately knew her type. He was a dapper gentleman with warm, downturned eyes; a Gallic nose; and a wide smile that likely appealed to the brash American couple. The man listened to Evalyn's lament: she, an enchanted newlywed, had searched far and wide for her wedding present but could find nothing, in all of Europe, to fit the bill. Whatever was she to do?

In Turkey, she'd visited Sultan Abdul Hamid II and appraised his harem, and nowhere had she seen jewels as beautiful as the ones worn by the exotic women in that far-off land. With a twinkle in his eye, the man at Cartier patiently listened. Then, when Evalyn finished, he uttered the words that sit at the tip of every shop clerk's tongue:

"We have just the thing for you," he said.

It was Cartier, after all, so maybe Evalyn McLean believed him. They were nearing the end of their honeymoon, so it's possible she would have gone for any old jewel just to have something shiny to bring home. But when the man returned, he wasn't carrying just any old jewel. In her memoir *Queen of Diamonds*, McLean describes the item: "A line of diamond fire in the square links of platinum where it would touch my throat became a triple loop and from the bottom circle was depended an entrancing pearl. It was the size of my little finger end and weighed 32½ grains. The pearl was but the supporting slave of another thing I craved at sight—an emerald. Some lapidary had shaped it with six sides so as to amplify, or to find at least, every trace of color. It weighed 34½ carats. This green jewel, in turn, was just the object supporting the Star of the East. This stone, a pear-shaped brilliant, was one of the most famous in the world—92½ carats. All lapidaries know it."

The price was six hundred thousand francs, or $120,000 (almost $3 million today). Even Ned, a noted spendthrift, balked. Evalyn promised she'd save them money by smuggling it through customs to avoid the considerable tariff. He relented. The Star of the East was hers.

When Evalyn brought the necklace home, Thomas Walsh found his daughter's recklessness funny. The McLeans, who'd intended their $100,000 gift to go quite a bit further than the shops, car dealerships, and hotels of Europe, were less amused. Either way, Evalyn adored her new diamond and found it surprisingly versatile. Every time she wore it, she fondly remembered her lovely, loopy honeymoon.

Pierre Cartier—the man in the store with the laughing eyes—remembered her, too.

◇◇◇◇◇◇◇◇◇◇◇◇◇◇◇◇◇◇◇◇◇◇◇◇◇◇◇◇◇◇◇◇◇◇◇◇

In Evalyn Walsh McLean's time, socialites were the real celebrities; it was their unpredictable lives and outrageous getups that got splashed across the gossip pages, rather than the wild stories about actresses, rock stars, and models that command page views today. America was fascinated by the inner workings of the "diamond horseshoe": the well-heeled and gem-spangled ticket holders at the Metropolitan Opera House, named for their constant presence in the coveted boxes of the theater's curved parterre. These families— the Morgans, Astors, Rockefellers, and Vanderbilts, to name a few— were the ones who helped build Cartier into a global brand. By the time the shop moved to the Rue de la Paix, Alfred Cartier and his three ambitious sons were contemplating other, bigger changes. International expansion was imminent; they just had to pick the right cities.

When Cartier first opened its doors in 1847, it sold work by other established jewelers like Fabergé, a Frenchman transplanted to Russia, where he delighted the Romanovs with his intricate and extravagant baubles, most famously his imperial Easter eggs. But over time, Cartier started carrying its own designs, too. The house experimented with platinum: a dense, silvery metal that had long flummoxed jewelers because of its extremely high melting point.

Cartier was one of the first manufacturers to master it, and to recognize it as a most suitable complement to diamonds. For ages,
silver had been considered the best match for white jewels because
it highlights blue tones and thus diminishes traces of yellow (put
another way, blue neutralizes any less-than-ideal amber or ocher
shades; that is why even now sellers present diamonds against a
backdrop of blue velvet). But silver is fragile. Besides tarnishing,
it also bends easily, making the jeweler's job less stressful but also
potentially threatening the longevity of a finished piece. Once jewelers figured out how to work with platinum, its one considerable
drawback—its lack of pliability—became its biggest selling point.
Unlike silver, platinum is all but indestructible after it is set.

Creating tantalizing tiaras, necklaces, bracelets, earrings, and
brooches, oftentimes influenced by the old Versailles style and the
jewels of Empress Eugénie, Cartier gained a following with the maharajas and maharanis in the East and the royals in the West. By
1902, a move to London—where the recently crowned king, Edward
VII, and his queen, Alexandra, were already Cartier's enthusiastic
supporters—was an obvious next step. Jacques, the youngest and
quietest of the three Cartier brothers, who expressed an interest in
travel, was chosen to run the new business, with the understanding that Louis, the oldest and most artistically gifted of the trio,
would stay in Paris and work alongside their father. That left Pierre.
In 1904, Louis and Pierre traveled to Russia, where they considered opening a third store in Saint Petersburg, but ultimately, they
settled on twice-yearly visits to the court of Nicholas II and made
the decision to concentrate their commercial energies elsewhere.
By 1907, Pierre had met and fallen for the American heiress Elma
Rumsey, originally of Saint Louis, who stood to inherit her industrialist father's fortune (when the press announced that she was

engaged to marry a well-off foreigner, the only surprise was that he wasn't titled). The intercontinental marriage settled it: Pierre Cartier was moving to New York.

Pierre, then thirty years old, was confident, affable, and shrewd, a combination that appealed to the wealthy Americans who became his peers when he married Elma Rumsey. He already knew J. P. Morgan, the steel baron and original investor in General Electric, who, as a husband and the father of three daughters (as well as one son), was a terrific Cartier customer. As opposed to Pierre's brother Louis—who cultivated an old-world European distinction with his waxed mustache, top hat, and cane; married a Hungarian countess; and courted a foothold within the European aristocracy—Pierre felt at home with the breezier, less formal Americans. By the time he met the McLeans in Paris and sold them the Star of the East, he was already preparing to open the New York location. They were exactly the kind of people he wanted on Cartier's US client list: rich, profligate, and notorious.

Pierre instinctively grasped the power of the press, especially in media-dominated cities like Washington, DC, and New York. That was likely an important reason why, in 1910—just a year after he opened the Manhattan store at 712 Fifth Avenue, and with the company freshly stretched across three countries—Cartier jumped at the opportunity to buy an infamous diamond called the Hope. A Parisian gem dealer had recently acquired the magnificent blue stone, a 45.52-carat cushion-cut brilliant. The price was $110,000—roughly $2.7 million in today's dollars—and the investment was risky, given that colored diamonds weren't in particularly high demand at the time. But the dark jewel had a fascinating lineage and already held a place within the collective imagination. Pierre Cartier was convinced he could sell it.

In fact, this particular diamond had already received a recent string of press. In November 1909, the *New York Times* reported that, after colliding with another boat on a foggy early morning, a French steamer called *La Seyne* sank fourteen miles off the coast of Singapore, killing 101 people. Among the victims was a Turkish gem merchant named Selim Habib. It was noted that, at the time of the wreck, Habib had a certain infamous blue diamond—the same one that Cartier purchased in Paris a year later—with him.

Like an undisciplined socialite, the Hope Diamond seemed to invite all kinds of rumors. Perhaps its bewitching color— somewhere in between a deep cornflower and steel blue—had something to do with it. But the incredible thing about the *Times* report is not just that it got the facts wrong—the Hope wasn't at the bottom of the ocean—but that the author managed to imply that the stone had caused the accident. "This adds to the list of misfortunes associated with the ownership of the famous gem," the article stated before launching into a summary of those misfortunes.

Habib had owned the stone; the paper was right about that. But he'd also sold it well before stepping foot on *La Seyne*. For the sake of the Hope's reputation, it wasn't actually critical that the stone was on the ship—the far more sensational detail was that Habib had died an unpleasant, untimely death. This was consistent with the narrative slowly curling itself around the Hope: the jewel destroyed its owners. The fact that it was spared, that it hadn't gone down with the ship, only made it seem more malevolent.

So why would Pierre Cartier risk touching the Hope Diamond? Quite simply, he didn't believe in the curse. His family had worked with diamonds for three generations, and in his experience they'd brought only good fortune. But he was a born salesman, and thought he could use the rumors to his advantage. A pitch had begun spinning through his mind like a thumping, whirling tune, and he had just the client in mind.

∞∞∞∞∞∞∞∞∞∞∞∞∞∞∞∞∞∞∞∞∞∞∞∞∞∞∞

A few years had passed since the McLeans' decadent honeymoon. Life carried on, shaped by their money in some ways and in other ways oddly untouched by it. In December 1909, Evalyn gave birth to their first son, Vinson, who slept in a golden baby crib—a gift from King Leopold of Belgium, a family friend. The press dubbed Vinson the "hundred-million-dollar baby," and Evalyn delighted in her new role as caregiver, thrilled by the chance to coddle the baby in the same way that she and Ned had been pampered. Unfortunately, the glow of new motherhood was instantly dulled when she learned that her beloved father was sick. Just a few months after Vinson's birth, Thomas Walsh died from an aggressive cancer that even the best doctors in the country couldn't treat. He willed half his fortune to his widow and the other half to Evalyn. In an effort to clear her head following the funeral and to rouse her mother from listlessness, Evalyn left her young son in the care of his grandmother, and she and Ned set off for Paris.

They stayed at the Hotel Bristol, a luxury accommodation on the trendy Rue du Faubourg Saint-Honoré. One morning while Ned, who loved to drink, was still hungover in his bathrobe and choking down a plate of room service eggs, a man came to call on them. Pierre Cartier was dressed in a silk hat, pressed trousers, and spotless cream-colored spats: an impeccable getup, which Evalyn McLean registered as a compliment. He was carrying a package and took a seat, even though Ned glared at him over his breakfast with crusty, red-rimmed eyes.

Cartier tapped the package with his fingers but otherwise didn't acknowledge it as he and Evalyn made small talk. Then, out of nowhere, he brought up the Turkish Revolution.

Evalyn reminded him that she and Ned had visited Constantinople during their honeymoon, when the revolution was already under way.

"Ah, I do not forget such things," Cartier answered. "You told me when you bought from me your wedding present, the Star of the East. I remember very well. It seems to me you told me then that you had seen a jewel in the harem, a great blue stone that rested against the throat of the sultan's favorite. A lovely throat, eh?"

Evalyn played along, though she couldn't quite recall a blue stone—still, she had no trouble believing that if she had seen one, she would have prattled on and on about it.

Cartier continued: "We hear the woman who had that jewel from the sultan's hand was stabbed to death."

Now he had Evalyn McLean's attention. This was no typical hard sell. Seeing he had his audience hooked, Pierre Cartier launched into his story. He kept the package tightly wrapped but close by, well within Evalyn's line of sight.

In Pierre Cartier's telling, the Hope Diamond originated in India, where it rested in the forehead of a Hindu idol. Jean-Baptiste Tavernier, a gem merchant and envoy to Louis XIV, spotted the jewel in his travels and plucked it from the idol's eye; knowing the French king's avarice for rare gemstones and his exotic taste, Tavernier believed the blue stone would command a fortune. Sure enough, Louis XIV loved the stone, and the Tavernier Blue became a centerpiece of the crown jewels. After the king died—of a painful, gangrenous hunting wound—the blue diamond remained the property of the royal family, and eventually it was worn by the capricious young queen Marie Antoinette. Suffice it to say, her story didn't turn out well. As for Tavernier: Louis XIV revoked the Edict of Nantes, which protected non-Catholics, and the Protestant gem merchant was banished, left to live penniless in Switzerland and eventually eaten by wild dogs. According to Cartier, this terrible fate was inscribed the moment Tavernier swiped the stone. He'd offended the Hindu god by stealing from him, and he and any future custodians of the haunted gem were condemned to pay for his mistake.

After the French Revolution, the Tavernier Blue disappeared, only to turn up fifty years later with a well-known dealer in London. Around 1830, the third-generation Anglo-Dutch banker and gem collector Henry Philip Hope, an heir of the illustrious Hope & Company, bought the stone and gave it his name. Hope survived his tenure of guardianship relatively unscathed but died childless, leaving no provision in his will for the future of the diamond. This left his three nephews to battle it out in court for a period of ten years, during which time the family was torn apart by greed and public antagonism. The jewel stayed in the Hope family until Lord Henry Francis Pelham-Clinton-Hope, a playboy descendant of Henry Philip whose movie-star wife left him for another man, sold it in 1901 in a desperate period of bankruptcy. Selim Habib, now resting in a watery grave, owned the Hope Diamond for a brief time after that. It's likely that he worked as a go-between for Sultan Abdul Hamid II, which is how the dazzling blue gem ended up dangling from the neck of his favorite concubine, but—*well*, we know how that worked out for her.

Evalyn Walsh McLean was rapt. Part of her recognized that Cartier was putting on an act, but what an engaging act it was! Pierre Cartier had spun his tale with all the conviction of a vaudevillian.

This strange peddler was far better company that her husband, who was still losing a fight against his eggs. She gave in to the moment. "Let me see the thing," she said.

Cartier didn't break character. Solemnly, he reached for the parcel. He ran a finger under a wax seal and unfurled the wrapping very slowly. Evalyn sucked in her breath. The stone was mesmerizing, as advertised, deep blue like a sapphire but with the hardness of a diamond. Its color was like the ocean from some angles, but like a stormy sky from others. Cartier reiterated that the stone was bad luck.

"Bad-luck items for me are lucky," Evalyn replied smartly.

Ned sensed a flirtation about to be consummated. "How much?" he piped up.

But before Cartier could answer, Evalyn interrupted. She didn't want it, she insisted; she didn't like the setting. There would be no deal. The spell was broken.

<hr />

Pierre Cartier left the Hotel Bristol disappointed but not dissuaded. After all, he had noted the look in Evalyn McLean's eyes when she admired the Hope—as an expert, he knew the gaze of the besotted when he saw it. He returned with the diamond to New York, where he promptly reset it in a more contemporary corona of brilliants and platinum so that the jewel could be worn either in a bandeau—a popular style of headband worn across the forehead or farther back as a hair accessory—or as a necklace. In November 1910, he sent the McLeans a letter at their Washington, DC, residence that announced his arrival back in New York and requested an appointment. Ned responded by phone, and that's when Pierre made a second, winning pitch: the stone was clearly meant for Evalyn. Would she like to take it for the weekend and mull it over, without any commitment to buy?

Evalyn took custody of the Hope, and what started on Saturday as mere curiosity became, by Monday, an unequivocal desire to own the jewel. Perhaps Cartier had been right and she'd wanted it all along, or perhaps the demon got its hooks into her. She told her mother-in-law she was going to buy it. The elder Mrs. McLean, who didn't love diamonds the way her daughter-in-law did, was aghast, citing the curse as a good reason to walk away from the deal.

What followed played out publicly in the press. In her autobiography Evalyn McLean claims that she tried to give the diamond back, but Pierre Cartier refused to take it. On January 29, 1911, the *New York Times* reported that the McLeans bought the stone for $300,000 (approximately $7.3 million today), detailing their elaborate plans to keep the Hope safe, which involved hiring two private detectives and buying a dedicated car to shuttle it to and from a safe-deposit vault. The article also summarized the "sinister history" of the "hoodoo diamond," which uncannily echoed the story Cartier told the McLeans in Paris, including the part where Tavernier was eaten by wild dogs. However, just a short while later, on March 9, the same paper announced that Cartier had filed suit against the McLeans for default of payment, claiming that he hadn't received a cent of the agreed-upon price of $180,000 (not $300,000). This article also specified a payment plan; allegedly, the Washington, DC, couple had consented to a $40,000 ($973,000 today) down payment, as well as a trade-in of a pendant worth $26,000 ($632,000) from Evalyn's collection. (The pendant was later identified as the very emerald-and-pearl one that supported the Star of the East.) Yet other sources, quoted in the *New York Times*, upheld the McLeans' story: "Friends of

the McLeans say to-night that there has been no transfer of ownership of the diamond; that the McLeans simply took it for inspection,
and have several times tried in vain to induce the Cartiers to take it
back." All the while, the paper treated the legal back-and-forth as yet
another example of the Hope's malign influence. On March 10, the
Times produced a copy of the contract between Maison Cartier and
the McLeans, supposedly signed on January 28 before Pierre and
Ned shared a champagne toast in the offices of the *Washington Post*. It
enumerated the financial terms of the deal ($40,000 down payment;
the brooch worth $26,000; plus $114,000 to be paid in three install-
ments) but also contained this eyebrow-raising clause: "Should any
fatality occur to the family of Edward B. McLean within six months,
the said Hope diamond is agreed to be exchanged for jewelry of equal
value."

The disagreement wasn't settled until February of the following
year, when the McLeans relented and bought the Hope for $180,000
(roughly $4.2 million today). It's unclear if the wealthy young
couple resisted payment as part of a prolonged negotiating strategy
or if the whole thing was a calculated publicity stunt. Either way,
the long-lasting lawsuit had the effect of fixing the idea of a Hope
Diamond curse in the minds of the public. As the fight between
Cartier and the McLeans unfolded, one expert even tried to rein
in the conversation about the stone with a levelheaded editorial.
T. Edgar Willson, then editor of the *Jewelers' Circular-Weekly*, wrote:
"Although the writer has been connected with the jewelry trade
press for nearly twenty years and has been especially interested in
the history of the great diamonds of the world, he never heard of
any stories connected with the Hope diamond until a sensational
article appeared a short time after it was brought to this country
in 1901. . . . It has been the custom not only to revive these stories

every time mention of the stone appears in public press, but to add to them fictitious incidents of misfortune."

Perhaps this clarification was a comfort to Evalyn Walsh McLean, who, as the new owner of the Hope Diamond, had the most to lose at the hands of a wrathful Hindu god and his powerful stone. For her part, she generally had a lighthearted attitude toward the curse, like a teenager willfully determined to scare herself with a Ouija board. McLean wore the Hope frequently—but supposedly refused to let any of her friends or family touch it. Not long after buying it, she had a priest come to bless it just to cover her bases. That way, in her words, she could let "the curse and the blessing fight it out together."

∞∞∞∞∞∞∞∞∞∞∞∞∞∞∞∞∞∞∞∞∞∞∞∞∞∞∞∞∞∞

So where did the rumors of a Hope Diamond curse come from? Pierre Cartier simply borrowed an existing narrative and ran with it. The real story of the Hope is less occult, but somehow more complicated and mysterious than the string of articles from the early twentieth century imply. In fact, the history of the stone is so poorly documented that experts and scholars are still, to this day, debating it. What's known for certain is that it first appeared in an 1839 catalogue commissioned by Henry Philip Hope to document his extensive gemstone collection. After Henry Philip died that same year, the diamond was purchased from the estate by his eldest nephew, Henry Thomas Hope, who subsequently willed it to his wife. She kept it until her death in 1884, when her will stipulated its inheritance by her grandson Lord Henry Francis Pelham-Clinton with two qualifiers: that he append "Hope" to his name, and that the diamond would be his only according to the provisions of a life estate, meaning that he kept it until his death and then ownership transferred to another living family member. Lord Francis agreed to both counts. However, in 1899—after years of gambling and financial mismanagement left him broke—he petitioned the court for permission to sell the Hope. The family fought back and he lost the appeal. But Lord Francis persisted, and in 1901, two legal go-arounds later, he found a buyer and the blue diamond forever left the family that had given it its name.

For a time, the diamond was in the possession of the New York jeweler Simon Frankel of Joseph Frankel's Sons, who then sold it to Selim Habib. In 1909, before the wreck of *La Seyne*, Habib sold the diamond to a Parisian dealer, who made the subsequent deal with Cartier.

Everything else that is generally believed to be true about the Hope Diamond comes from a mix of historical detective work, forensics, and informed speculation. It's widely accepted that the Hope's mother stone is the roughly 112-carat *beau violet* (beautiful violet) described by Tavernier in *Les Six Voyages de Jean-Baptiste Tavernier*, which details the author's six gemstone-buying trips to India. Unlike other jewels Tavernier describes in the book, the *beau violet* (a.k.a. the Tavernier Blue) doesn't have an origin story, leaving room for conjecture that it was purchased under suspicious circumstances—or, more fancifully, stolen from a Hindu temple under cover of night. The latter seems to be Cartier's unique embellishment. When Tavernier arrived back in France, he sold the blue to King Louis XIV, who held on to it for a few years before having it recut into a $67\frac{1}{4}$-carat brilliant, in keeping with the current style to sacrifice weight for beauty.

King Louis's heart-shaped diamond was called the Blue Diamond of the Crown and has come to be known colloquially as the

French Blue. Louis XIV loved to wear his jewels, and the blue made frequent appearances over the years. Then his successor, Louis XV, was appointed to the Order of the Golden Fleece, a brotherhood devoted to the protection of the Catholic Church. The king had a special insignia designed to commemorate this honor, which he called the Toison d'Or, or Golden Fleece: a spectacular hanging jewel that featured rare yellow sapphires, hundreds of brilliants, and three very large colored diamonds, including the French Blue. Louis XVI inherited the Golden Fleece intact, and gemstone historian Ian Balfour cites this as an argument against the possibility of Marie Antoinette's having ever worn the blue, as the Golden Fleece was an "exclusively male ornament."

At the first whiff of revolution, the crown jewels were moved into the Garde Meuble, a temporary museum with minimal security that was promptly robbed. The French Blue disappeared, leaving a significant and tantalizing gap in its history, further complicated by the fact that the Hope Diamond is about thirteen carats lighter than Louis XIV's stone. Experts have convincingly (and scientifically) proved that the French Blue and the Hope are one and the same, but not before floating some wild theories about what happened in the interim (the most extreme has it playing a pivotal role in the French Revolution). More likely, whoever had the stone kept mum until the French government's twenty-year statute of limitations on the theft passed. A plausible explanation for its weight loss, and one that has been floated by a number of diamond historians, is a deliberate effort to disguise the French Blue and make it all the more difficult to trace back to the robbery in Paris.

Eventually, the blue diamond landed in the collection of Henry Philip Hope. How it got there is—again—a mystery, though one fun theory places his uncle in the court of George IV of England when a gem dealer tried to sell it to the king. The many question marks in the history of the Hope make it tempting to fill in the blanks with tales of intrigue and disaster, especially given the high-profile cast of characters. But the truth is that up through Lord Francis's tenure with the stone, there's no evidence to support a curse. Tavernier died at the age of eighty-four in Moscow, without a wild dog in sight. Even Lord Francis, who suffered a few significant blows, seems to be a victim of his own bad choices.

One could also make the argument that any object—a diamond, a vase, a silver candelabra—that sticks around for hundreds of years is apt to witness a fair share of misfortune; the possibility gets amplified if that same object happens to be mixed up in issues of greed and power. But rational explanations aside, there's no doubt that the curse of the Hope Diamond makes for a good story, and that early-twentieth-century people loved reading about it. The Hope was a mystical gem that entranced the rich, only to destroy them. At a moment of terrific economic inequality, what could be more gratifying than that?

<hr />

After settling the lawsuit with Cartier and getting the diamond blessed, Evalyn Walsh McLean wasn't terribly worried. Yes, she received letters almost daily telling her to run from the ruinous stone, including a particularly amusing one from May Yohe, the actress who'd been married to Lord Francis and apparently blamed the Hope for the misery in her life. She begged Evalyn to destroy it. But the giddy young socialite did nothing of the sort. She debuted her new plaything at a party she hosted for the ambassador to Russia, George Bakhmeteff, a family friend of the McLeans'. Guests knew to keep an eye out for it, as the *Times* had already reported that she intended to wear it that night. Once again, the media took the opportunity to share details of the couple's extravagant lifestyle: the $8,000 ($191,000 today) in English yellow lilies they'd imported for the decor; the estimated $30,000 ($716,000) they planned to spend on food, drink, and music to entertain a roster of just fifty prominent guests.

As promised, Evalyn McLean wore the Hope as a necklace, along with the Star of the East in her hair. (She maintained a more-is-more attitude when it came to wearing her jewels.) The Hope became a signature accessory, which she wore to the decadent, champagne-fueled parties that dotted her social life. In one particularly colorful anecdote, the McLeans rang in the New Year with another lavish party at their DC mansion. Supposedly, Evalyn set the tone for the night by greeting the guests wearing the Hope Diamond—and little else.

And yet—perhaps to nobody's surprise except her own—life with the blue gemstone wasn't just flowers and cocktails. If any one particular custodian could speak to the authenticity of a Hope Diamond curse, it would be Evalyn. The McLeans finalized the deal with Cartier in February of 1912; by September, Ned's beloved mother, who'd railed against the purchase, was dead. In her memoir Evalyn remembers, "I hated my Hope diamond" when Emily McLean passed. But that loss was just a momentary sting compared with the one that came next, in 1917, when her eldest son, Vinson—now a nine-year-old brother to two younger boys, Ned and Jock—was hit by a car in front of their house. At first, he seemed shaken but fine. By afternoon, his condition had taken a turn for the worse and he was paralyzed. By nightfall, he was dead. In Evalyn's words: "What happened to us all was just about as tragic as if each one, instead of only I, had worn a talisman of evil." In some ways, a curse seems less cruel an explanation than the chilly randomness of life.

For a long time, Evalyn's husband, Ned, had teetered on the precipice between hard drinking and full-blown alcoholism, and after Vinson's death, he dropped willfully into the abyss. There were times when Evalyn tried to intervene; at other moments, it was easier to ignore the situation entirely. But in 1928, after the McLeans had one more child, a daughter, they separated. The pair who had spent their whole lives together had to muster the strength to rebuild, but Ned couldn't. He drove the *Washington Post* into bankruptcy and drank himself insane. In 1933, following a doctor's warning that he'd probably wind up homeless, Evalyn had him committed to the same sanatorium where Zelda Fitzgerald was held. Evalyn visited frequently, but Ned wouldn't speak to her. When he died, he left her out of his will completely, though by that time he had accrued so much debt that Evalyn—who never finalized their divorce—was left responsible for it. At one point in the early 1930s, her finances were so drained that she drove to New York and hocked the Hope Diamond for $37,500 (roughly $598,000 now).

The moment she felt flush again, she bought it back.

The final blow came much later, when Evalyn's daughter, Evalyn "Evie" McLean Reynolds—the wife of a much older US senator and mother of a baby girl—died of an apparent overdose of sleeping pills. Evalyn followed soon after, succumbing to pneumonia at the age of sixty-one.

∞∞∞∞∞∞∞∞∞∞∞∞∞∞∞∞∞∞∞∞∞∞∞∞∞∞∞∞∞∞∞

For anyone inclined to argue in favor of the Hope Diamond curse, Evalyn Walsh McLean's story provides healthy fodder. But as her memoir shows, she herself never blamed the jewel, except in a few fleeting, vulnerable moments. Admirably, she didn't hide behind superstitions, but instead took responsibility for the role that money—and not the things it bought—played in her life. For her and Ned, money was like a drug: it made them lazy and selfish, and although its effects could be euphoric, they were always, ultimately, temporary. In her later life, sobered by loss and financial troubles, Evalyn tried to make good by supporting veterans' causes, writing a fairly serious column for the *Washington Times Herald*, and even attempting to help out with the Lindbergh kidnapping case, paying her own money to con man Gaston Means for ransom.

And anyway, to blame her diamonds would be to deny herself a simple pleasure in a life full of considerable pain. As she put it: "It is no use for anyone to chide me about loving jewels. . . . They make me feel comfortable, and even happy."

5

The Optimists

New York

I f there was anyone who understood the almost rapturous appeal of a beautiful diamond, it was the self-made American wholesaler, dealer, and jeweler Harry Winston. The son of a New York jeweler, he'd honed his eye for gemstones since boyhood, and by the time he grew into a veritable force in the industry—a short man in a black homburg hat who conspicuously wore no jewelry—a natural aptitude had developed into a full-fledged obsession. When Winston sold a unique, expensive diamond, he didn't celebrate like other dealers; he pouted. For him, giving up a diamond—especially one that he'd personally transformed from rough into a perfect polished stone—was bittersweet no matter the final price, like walking a beloved daughter down the aisle. In fact, that's how he talked about his diamonds, as if they were children. This one was his "little fella"; that one was his "baby."

He never believed that the women who wore them truly appreciated their magic. "Adornment!" he scoffed in an article for *The New Yorker*. "They'd wear diamonds on their ankles if it was stylish! They'd wear them in their noses! They have no feelings for diamonds." This attitude—Winston's inability to emotionally disengage from his products—initially drove his wife crazy. In fact, it almost derailed their courtship. Winston met the slim and cheerful Edna on a train to Atlantic City, after she'd had her tonsils removed and her father took her to the beach to recuperate. Their connection was immediately apparent, but the more she got to know him, the more she realized that he reserved a part of his heart for a mistress: his jewels. Not convinced she'd be happy with this arrangement, she accepted an offer of engagement from a simpler man. Harry Winston called her up two days before the wedding.

"Why don't you marry me instead?"

In the end, Edna couldn't deny her true feelings and she married him in New York the next day. They spent their honeymoon talking about diamonds. Loving Harry Winston meant accepting that his focus would always be split between work and family. His stones really were his other sons and daughters. Maybe Edna could make peace with the fact that her husband had many, many offspring—two flesh and blood, countless others sparkling—as long as she was his one and only wife.

Besides, Winston's total preoccupation with diamonds was what made him so successful. There was nothing in his history that suggested he'd one day become a force in the international gemstone business. He was born in 1896, the third son of a jeweler whose family lived on the third floor of a walk-up apartment building on West 106th Street in New York City: a tree-lined working-class neighborhood made up of geometric row houses teeming with Jewish and Italian immigrants. Winston's father owned a local jewelry store on Columbus Avenue, and Harry—a small, puckish boy—liked to accompany him to work. After Harry's mother passed away, his asthmatic father moved the family to Los Angeles for the hot desert air, and opened a new store on Figueroa Street. One day, at just twelve years old, Harry paused at a display of costume jewelry in a local pawnshop window, under a sign that advertised "25 cents: Take Your Pick." A single green stone stood out, scintillating in the sunlight. He bought it, and brought it to his astonished father, who confirmed Harry's suspicion that it was an emerald. As the story goes, the not-quite-teenage Winston sold it two days later for a profit of $799.75 (about $19,800 today).

He was hooked. He stayed in school until he was fifteen and then dropped out to join the family business. His father admired his natural talent but also recognized certain warning signs that concerned him. "My father was always afraid that jewels would someday possess me," Winston told *The New Yorker* in 1954. "He used to say to me, 'Harry, you're the master of your jewels now, but . . . someday [if you're not careful], your jewels will master you.' " After a few years in Los Angeles, Winston returned home to New York. Although the West was flooded with oil wealth, the East Coast was, indisputably, the center of the American industry. He rented an office at 535 Fifth Avenue, around the corner from the new diamond district, which had recently begun a migration uptown from picturesque Maiden Lane. On Forty-Seventh Street, jewelers had easier access to the trains at Grand Central Terminal and Pennsylvania Station, and were walking distance from the hotels and department stores that had also converged in the city's midtown, turning what was once a neighborhood of high-society residences into a bustling commercial corridor. As soon as a few respected jewelers made the move, others felt compelled to follow. Winston's new office positioned him perfectly to take in the action; to absorb the lingo of the diamond dealers who tempted one another with "Jager" diamonds, from the Jagersfontein mine, said to have the clearest blue-white color; and to make social connections of his own. He called his early venture the Premier Diamond Company.

By this time, diamond wealth was moving away from Europe and toward the United States. As the First World War broke out and then spread, the overseas demand for gemstones plummeted so quickly that Americans, keeping their eyes on the action from behind their morning newspapers, became the only consumers who still had money to spend on diamonds. European dealers responded by shipping their stock across the Atlantic. Luxuries of all kinds changed hands with the frenzy usually reserved for the stalls of an outdoor bazaar. The stores on the Rue de la Paix shuttered; the streets no longer felt stylish with bombs raining down on France.

The Europeans had more pressing concerns than jewels, and this suited wealthy Americans just fine; they were more than willing to buy up any stones that previous owners—whether out of pragmatism or desperation—had to release into the marketplace. But then the diamond mines closed. As the fighting trickled down to South Africa, where Great Britain and Germany both upheld a colonial presence, De Beers rightly recognized that the global appetite for diamonds had been curbed by bloodshed in almost every country but one. In 1916, an editorial writer for the *New York Times* pointed out the absurdity of Americans clamoring for gemstones at a moment when the rest of the world was suffering: "The economic dictator of Germany increases the restraints upon food consumption. . . . The British Government by Order in Council puts prohibitions on the importation of luxuries, including even soap, and the Chairman of the Good and Welfare Committee of the

National Jewelers' Board of Trade in New York talks ominously of a diamond famine because the reopening of the mines of South Africa has
been postponed and the demand for gems at the same time is increasing. What a crazy world! Not enough food in some places and not enough
diamonds in others."

However, Americans could be scolded, but they would not be
shamed. With the impression of scarcity came higher diamond prices,
especially for the smaller stones used in elaborate jewelry settings. By
1919, the reported price of an average-quality stone hovered around
$300 a carat (or $4,240 today), while small diamonds, called melee, cost
$450 to $500 (roughly $6,700) for a bundle of the same weight. While
this seemed counterintuitive, it actually made sense given that the
smaller stones required more precise, expert fashioning, and two of
the three major cutting centers—Antwerp, Amsterdam, and New York—
had all but gone dark. When the National Retail Jewelers' convention
took place in Chicago in summer 1919, president Joseph Mazer predicted a solid future for diamonds, citing record-breaking paychecks
and even Prohibition, which in his estimation would leave working-
men with more money to spend on their wives and daughters. Flappers,
with their loose, shapeless dresses and short haircuts that showed off
the ears and neck, loved jewelry—the more sparkle they brought to dim
nightclubs, the better.

The stateside industry had good reason to be bullish. But then
American jewelers couldn't possibly see the rogue wave roiling up
ahead.

∞∞∞∞∞∞∞∞∞∞∞∞∞∞∞∞∞∞∞∞∞∞∞∞∞∞∞

Harry Winston was undeniably bold, but as a newcomer he had significant challenges to overcome, even given the hospitable climate

for diamonds. He was tiny—estimates range from five foot one to five foot four—with a round face and almost elfin features. When, at twenty-four years old, he arrived at the New Netherlands Bank of New York for a meeting to discuss his loan, he was promptly dismissed before he could explain who he was and told to go fetch his boss. Winston took the hint and hired a white-haired, distinguished "face" of the business—a decision that presaged his future life in the shadows when, after he'd inherited the mantle of "King of Diamonds," his insurance company, Lloyd's of London, prohibited him from being photographed, warning against a potential kidnapping. Unfortunately, Winston's youthful appearance wasn't the only hurdle. Despite certain setbacks having to do with the war, De Beers still owned over 90 percent of the world's diamonds, and without the right contacts and a venerable name, it was almost impossible to buy them at anything close to wholesale. As expected, De Beers wasn't particularly eager to work with young Americans who fancied themselves the next Tiffany. Harry Winston loved diamonds more than food—he was known to survive on just coffee with saccharin during business hours—but he couldn't figure out how to get his hands on the good ones.

The answer came when he looked to auction and estate sales, which were filled with high-quality, if antiquated, jewels that, given the right skills, could be recut or reset into more appealingly modern pieces. The great collectors—the women of the diamond horseshoe, Cornelia Bradley-Martin's generation—were passing away and leaving behind elaborate bequests that were too valuable for the families to keep, especially if there wasn't a female next in line for inheritance. That's exactly what happened with the great jewelry lover Arabella Huntington, who died in 1924 and left behind an estimated 1.275 million dollars' worth of baubles ($16.7 million today)—as well as a substantial income tax bill. Her only son, Archer, the executor of the estate, had no interest in diamonds, nor did his understated wife, Anna Hyatt, a San Francisco sculptor with no need or desire for old-fashioned tiaras and cocktails rings.

Yet for the right person, the collection was dazzling: beautiful, yes, but also a veritable journey through jewelry history. Over the course of her life, Huntington had accumulated the finest examples of craftsmanship from shops like Cartier and Tiffany & Co., as well as older, more obscure jewels purchased from her network of art dealers. Among her lifetime acquisitions were a diamond chain from downtown New York gemstone importer Alfred H. Smith and Co., with a hanging pendant made up of a pear-shaped natural pearl dangling from a seven-carat pear-shaped emerald; a diamond-and-black-pearl brooch from Tiffany & Co.; and a very famous string of rare black pearls, formerly owned by the Duchess of Hamilton, that Huntington had bought for a price of 100,000 francs. Her son and daughter-in-law were happy to sell almost the entire lot to the highest bidder—the budding jeweler Harry Winston. The investment required an enormous $1.5 million loan ($20.2 million today) from the New Netherlands Bank of New York.

Winston was just beginning to make a name for himself when the death knell of the luxury business sounded on October 29, 1929: Black Tuesday. No matter—it takes hubris to be the king. At least diamonds weren't pegged to the stock market. When the Depression hit, Winston joined the other diamond men on Forty-Seventh Street in weathering the crisis. If he was worried, he didn't let it show. In 1932, he shuttered the Premier Diamond Company—and founded Harry Winston Inc.

Not everyone in the industry managed to survive the financial meltdown. In 1929, two debonair brothers with a strong foundation in the jewelry industry and a flourishing flagship on Place Vendôme took a leap of faith and signed a lease for a New York City boutique. They were Louis and Julien Arpels, and their store was Van Cleef & Arpels, founded by their eldest brother and their brother-in-law around the turn of the century. The doors opened in New York on October 24, the first day of the Wall Street crash. They shut almost immediately afterward. The Arpels brothers accepted defeat, and for years, they gave up on the possibility of coming stateside.

Harry Winston, on the other hand, had gathered enough momentum to keep his business afloat. In fact, he thrived on the drama, calling the diamond business a "Cinderella world." The more showpieces and rare stones he discovered at estate sales, the more he became obsessed by the mission of buying up the best stones from around the globe. He was also a brilliant self-promoter, and as he sat in his New York City office, wondering how to set off sparks in the industry after the dullest days of the Depression, he remembered the way the world once went mad for huge diamonds like the Hope. The right jewel was so much more than a rock: it was a shimmering message that the earth could be generous and even magical. What better moment was there for a reminder that the darkest depths could eventually produce something gorgeous?

Then, with fortunate timing, the Jonker emerged.

Before the Jonker (pronounced *YON-ker*), the last huge diamond found in South Africa was the Cullinan: a 3,106-carat rock discovered in the Premier mine in 1905, the largest gem-quality rough

ever unearthed. This new stone was discovered by Johannes Jacobus Jonker, an old-fashioned prospector of Dutch ancestry who had
spent eighteen years hunting for *his* Cullinan: the stone that would
change his life. On the morning of January 17, 1934, the sixty-two-
year-old father of seven was bone-tired. It had been years of the
same, moving his family from claim to claim, always holding out
hope for the next big discovery, always wandering home—dusty and
hungry—wondering how he would feed his kids that night. A storm
had come through that morning, and Jonker knew the wet soil
would be heavy. The wind whipped. Some days, hope wasn't enough
to blunt the edges of poverty. Jonker decided to stay in bed.

But the family couldn't afford a day off, and he sent out his son
Gert and two native diggers. Later, they'd tell him how it happened:
Digger Johannes Makani was rinsing a bucket of rocks when he felt
something hard and unusual and his whole body tensed. Silently,
he scrubbed the egg-shaped stone until he got at its icy surface.
Then he tossed his hat in the air and announced:

"Good baas [master], I have found it!"

Gert ran home to show his father, and the old man fell to his
knees. At 726 carats, it was the fourth-largest diamond ever found.
That night, before the stone could be locked in a safe, Jonker's
wife wrapped it in a cloth and slept with it hanging from her neck,
while Jonker and his boys loaded their guns and took turns standing guard. By the next morning, news of its discovery had spread,
and Jonker made plans to take the blue-white stone to Johannesburg. There, the Jonker Diamond received its official name and
De Beers, on behalf of the Diamond Corporation Ltd.—the arm of
the company responsible for diamond distribution—bought it for
63,000 pounds (4 million pounds today, or $6.2 million). An elated
Jonker told the press that with his earnings, he planned to buy two

thousand acres of land in the Transvaal and start a farm. As for the
other Johannes, the employee who actually found the stone, Jonker
promised that he would have continued employment at the farm
and would "get a good present."

Almost immediately, diamond insiders speculated about the
Jonker. Would Joseph Asscher, who had famously fashioned the
Cullinan for the British crown, cut this stone, too? Since it was
found only five miles away from the original Cullinan, and had a
similar shape and color profile, some professionals wondered if it
was actually part of the same mother stone (this theory has since
been discredited). King George V's Royal Silver Jubilee was coming
up, celebrating the twenty-fifth anniversary of his coronation, and
a handful of experts—likely those close to the royal couple—noted
it would make the perfect gift for him and Queen Mary. Like the
Cullinan and the Hope, the Jonker captivated the public—as well as
Harry Winston, who was watching, gimlet-eyed, from the shores of
America.

He wasted no time. Quickly, he put out feelers through his network of diamond dealers: Would De Beers consider selling the
Jonker? Negotiations took months, during which time it reportedly cost the Diamond Corporation $15,000 a year in insurance
($265,000 today). On May 16, 1935, the *New York Times* announced,
"Famous Diamond Is Coming to U.S."—almost as if Harry Winston's
reported $730,000 purchase (a whopping $12.5 million) was an act
of selfless patriotism. Winston assured the press that he'd send it
home on an American ocean liner. When the time came, he kept
his word and sent it by post for 65 cents. Years later, after Winston
had made a practice of sending priceless gems by US mail, he explained it was the safest method, for him, his employees, and the
jewels themselves. Perhaps this was true—or maybe he noticed the

way the Cullinan made headlines when it traveled from Johannesburg to London the same way.

With the Jonker in his possession, Winston courted publicity at
every turn. When the stone arrived in New York, it wasn't delivered
to the offices of Harry Winston Inc., but instead made a carefully
orchestrated stop at the American Museum of Natural History on
Central Park West. The Jonker arrived uncut, which meant that
Winston could supervise the process of fashioning it himself, with
the additional benefit that he was spared the tariff since rough
gems weren't subject to the same duty as polished ones. On June
20, 1935, an armed guard marched an unassuming wooden box
up the museum steps and placed it on a pedestal, draped in black
velvet, in the Morgan Memorial Hall of Gems. It was signed for by a
female employee of Harry Winston's, a Miss Gladys B. Hannaford (a
woman who, many years later, would take a unique public relations
job with De Beers and earn herself the sobriquet "Diamond Lady").
The unveiling was deliberately ritualistic: a "secret ceremony" conducted by the museum director, enacted for fifty specially invited
members of the press. As cameras flashed, the director opened the
box and began unwrapping layers of cotton and tissue paper. When
the rough stone finally appeared, at least one journalist was struck,
amid so much pomp and circumstance, by the ordinariness of the
thing: "It looked for all the world like a piece of the camphor ice the
old-fashioned housewife used to keep on the bathroom shelf, and it
was about the size of one of the pieces of coal in that same housewife's kitchen coal scuttle."

Still—trompe l'oeil notwithstanding—Americans rushed to
see the "camphor ice" themselves. The museum reported that over
the next three days, more than five thousand people paid homage
to the Jonker, secured behind bulletproof glass, before Winston's

insurer ruled that even this arrangement posed too significant a risk. Anyway, a weekend was enough time to secure public investment in the stone, but just in case it wasn't, Harry Winston had one last trick up his sleeve. Once the Jonker was safe in his office, he put in a call to 20th Century-Fox and hired Shirley Temple to shoot some publicity photos with the uncut gem. At that time, America's ringlet-haired darling had just turned seven and had already charmed audiences singing "On the Good Ship Lollipop" in *Bright Eyes* and tap-dancing with Bill "Bojangles" Robinson in *The Little Colonel*. In the photos, Temple holds the diamond, which is almost the same size as her hand. She rests it against the black bow of her collar, in the middle of her chest, as if to say: "Be still, my glittering heart." In another, she lifts it up in the air like a paper airplane and looks at it askance, her eyes expressing everything Winston wanted the public to think: *Holy moly.*

With the rough Jonker properly commemorated, the next step was to fashion it. It had already been studied by experts in Europe who submitted their recommendations for how it should be cleaved; still, Winston wanted the work of a master, and so he hired Lazare Kaplan, a Belgian artisan who came to New York at the turn of the century and founded his own diamond business. Technically, Kaplan was Winston's competitor, but he couldn't resist the opportunity to show his son Leo, then in his early twenties, how to cut such a huge stone. The reported $30,000 ($522,000 today) Winston agreed to pay him probably didn't hurt either.

Years later, Kaplan admitted that he thought the job would be relatively straightforward, given the preliminary assessments from the European cutters. He'd follow their instructions and voilà: a little bit of press, an education for his son, and a serious paycheck. But it wasn't that simple. As soon as he sat down and truly examined the stone, he saw signs of trouble—a tiny ledge, like an active fault under the earth, that the other experts had overlooked. The European plan spelled disaster. More, Lloyd's of London refused to cover the cleaving process. Kaplan understood that Winston was gambling three-quarters of a million dollars on a cutter's expertise and nimble fingers.

For months, Kaplan analyzed the Jonker. He made a plaster model—then another, and another—until one day he realized that he had over a thousand. He made lead models, too, and marked them with black India ink, determining where he'd make cuts, what size chunks would result, and how to minimize carat loss while, at the same time, protecting the integrity of the stone. At one point he felt ready to cut it, going so far as to make a tiny incision with another diamond, before he noticed a curve in the rock he hadn't seen before and reconsidered. Finally—almost a year after he accepted the job—Lazare Kaplan hit the tipping point and realized additional strategizing would be a liability. In preparation to cut, he spent a full week etching grooves on the surface of the diamond. After so much time had passed, he could afford to be meticulous.

On Monday, April 27, 1936, with Kaplan and his son Leo preparing to finally cleave the Jonker, the press gathered at the offices of Harry Winston Inc., expecting to watch. That's what the woman on the phone had promised—journalists were invited to document the process. And yet, when they arrived, reporters and photographers alike were told they were barred from entering the room where the Kaplans were cutting the diamond. Not even the owner of the Jonker, Harry Winston himself, would be a witness that morning. What went on was a sacred rite between father, son, and stone.

Behind closed doors, Leo Kaplan positioned the rough in a clamp, as he had done with so many of the models. At twenty-three, he was already an accomplished artisan himself, and he and his father had cut countless other gemstones between them. If they could ignore the retinue of journalists outside—forget the dollar value of the Jonker for just long enough to drop the blade—their plan was infallible. They'd already made the first cut: a small marquise from the end. But this big cut was the one that counted—the one that, if executed properly, would take advantage of the natural grain and split the rock in two. When the diamond was secure, the younger Kaplan took a wedge to one of the grooves his father had traced on it. Lazare stood over him. He hit it with a quick, precise tap.

"It's perfect, Father!" Leo rejoiced as the Jonker fell cleanly in half.

The next day, the Kaplans reenacted their terrific success for the newspapers. In the photograph that ran in the *New York Times*, they both looked deep in concentration, but Leo's mouth was turned up slightly at the corners, as if he couldn't keep from smiling.

After that, the rough Jonker was transformed into twelve beautifully fashioned jewels: eleven emerald cuts ranging from a 3.5-carat baguette to a 14.3-carat diamond dubbed "Number One" and a 15.8-carat marquise, a bit sentimental because it had been the first. Of course, it wasn't as special as Number One, the prize of the collection, which, although smaller than King Edward VII's Star of Africa (cut from the original Cullinan), was said to be more perfect. Together, the twelve Jonkers had an overall estimated value of around $2 million ($33.7 million today), and when they went on sale, Winston, ever the protective papa, confessed his wish to find one buyer so that they could stay together. Whether or not this was realistic, one couldn't fault him or Lazare Kaplan for getting attached to the Jonker. Almost two decades after the diamond had been cleaved, polished, and sold, the pair bickered like divorced parents in *The New Yorker* about who really ushered it into being, with Kaplan griping that Winston took too much credit for his craftsmanship and Winston retorting that he had to recut ten of the twelve jewels himself.

True to form, Harry Winston put the finished Jonkers on display, again at New York's Museum of Natural History, where even more visitors came to see them than had turned out for the uncut stone. The run of the exhibit was only three days, and presumably, the twenty-four-hour patrol of armed guards, the steel vault, and the bulletproof glass protecting the diamonds were as much for the press's benefit as they were precautions required by Lloyd's of London. As breathtaking as the Number One Jonker was, Winston had many other opportunities to show it off over the next decade because it took more than ten years to find a buyer. In 1939, Number One—by then just called the Jonker—was the star of a jewelry show put on by Jazz Age jeweler Paul Flato at New York's Ritz-Carlton Hotel, where socialites modeled diamonds. It wasn't until 1949 that King Farouk of Egypt bought it. Dream as he might, Winston was unable to keep the collection together: other Jonkers were dispersed across the globe, rumored to have ended up in the hands of illustrious men like the maharaja of Indore and John D. Rockefeller Jr. As for Johannes Jacobus Jonker, his small fortune is said to have scattered like the stones that bore his name. He bought himself land, livestock, and even a limousine, but apparently squandered or lost the rest of his money. The life of a workaday farmer didn't suit him, and within a few years of finding and selling the Jonker Diamond, he was an anonymous itinerant digger yet again.

◊◊◊◊◊◊◊◊◊◊◊◊◊◊◊◊◊◊◊◊◊◊◊◊◊◊◊◊◊◊◊◊◊◊◊

Harry Winston wasn't the only one in the jewelry industry working to crawl out from the dark pit of the Depression. The lean years had

been hard on Tiffany & Co.; in 1933, when sales still hadn't recovered since Black Tuesday, the business was forced to lay off employees at all levels. The company was in a state of perpetual triage all
the way through 1939. Meanwhile, at Cartier, the New York branch
had begun issuing catalogues to keep up with the competition, and
its 1932 issue featured "Gifts one dollar upward" in deference to the
economic climate. Suffice to say, it was a terrible moment to work
for a luxury brand. So when, in 1937, Pierre Cartier heard of an opportunity to participate in an international event that was sure to
be highly publicized, he was intrigued. Consumers had been stuck
in a state of deprivation and panic for so long that they'd come to
think of jewels as impossibly extravagant—that is, if they thought of
them at all. It was time to remind them of all that diamonds could
be: uplifting, exotic, enchanting. He made a few calls.

The World's Fair of 1939 was the brainchild of New York City
power players who believed it would take something momentous to
remind citizens that what had come unglued could be pasted back
together again. In November 1937, the fair's organizers accepted
a joint application from competitors Cartier, Tiffany & Co., and
Marcus & Co.—a jewelry business founded by German immigrant
and ex-Tiffany employee Herman Marcus and his two sons—to collaborate on an exhibit. When the "Jewelry Group" balked at the
financial obligations of building their own structure, as many of
the fair's corporate participants had committed to do, organizers
suggested that they reach out to a company that might jump at the
chance for a bit of upbeat press: De Beers. Sure enough, the syndicate agreed to sponsor the event, and with further participation
from Udall & Ballou (a New York jeweler) and Black, Starr, Frost
& Gorham (a jeweler with roots going back to 1810), the House of
Jewels was born.

The 1939 World's Fair was the second-largest American world's fair after 1904's Louisiana Purchase Exposition in Saint Louis, and it took place at the sprawling Flushing Meadows Park in Queens. It was meant to evoke the greatness of America's past—it opened on April 30, the 150th anniversary of George Washington's inauguration—as well as the promise of brighter days ahead. The theme was "The World of Tomorrow," and if that wasn't enough to drive home the message, the fair's slogan was "Dawn of a New Day." Many of the exhibits, from companies like Ford, General Motors, and RCA, emphasized the linked notions of progress and innovation. Fair admission was 75 cents for adults and 25 cents for children, so for $2—that's $33.29 in 2014—a family of four could enjoy acres of attractions from ten a.m. to ten p.m. The House of Jewels was one of the smallest buildings, but it was centrally located. At the official opening of the exhibit on May 16, 1939, fair corporation president Grover Whalen thanked Pierre Cartier and then waxed poetic about the display: "Jewelry is something more than adornment. It is perhaps the most complete and most beautiful summary of the design and the art of its producing period."

Sure enough, fairgoers were drawn to the Jewelry Group's minimalist white facade, which in its simplicity stood out against other, more colorful buildings, and inside stored over $5 million ($83.2 million today) in precious gems. The tour started with De Beers's own impressive contribution: two million dollars' ($33.3 million) worth of rough and polished stones. As visitors entered in groups, quiet music played and a prerecorded voice narrated the story of diamonds. Choreographed spotlights flashed on and off, emphasizing different parts of the tale: the discovery of rough, its subsequent journey around the world. Afterward, to best illustrate the diamond's transition from glassy rock to coveted jewel, the five participating New York jewelers exhibited their own creations, some from recent collections and others made especially for the World's Fair. Tiffany & Co. staged a particularly impressive display. The major showpiece was the 128.51-carat cushion-cut Tiffany Diamond surrounded by a spray of 635 stones of varying shapes and weights totaling 362.64 carats and worth $200,000 ($3.3 million now). There was also an emerald-and-diamond tiara worth $110,000 ($1.8 million) and made of 906 gems, and an aquamarine-and-diamond necklace, which had three strands of round diamonds and a 217.57-carat aquamarine in the center. Rings, brooches, pearls, and a spread of gold and silver accessories rounded out the assortment. None of these items were available to buy at the World's Fair, but after it was over they were moved to the Fifty-Seventh Street store, where many of the items were purchased by wealthy customers.

The 1939 World's Fair was a popular success; from April to October, Flushing Meadows Park hosted twenty-six million paying customers. The turnout was also enough to justify reopening for the summer of 1940, with a repeat performance from the House of Jewels. The theme of the 1940 fair was "For Peace and Freedom." By that time, the war in Europe had mushroomed from an isolated set of conflicts into a mass of violence that had even prevented some European participants from returning home after the 1939 season. Among them were Louis and Julien Arpels; although their first experience in New York in 1929 had been discouraging, their time at Flushing Meadows was different. When the choice came to travel back to the Nazis or try again in the Big Apple, Louis and Julien, along with Julien's son Claude, decided to stay.

In 1942 Van Cleef & Arpels moved into its American flagship at 744 Fifth Avenue, situated on a lot once owned by railroad baron Collis Huntington, Arabella's husband. It was right across the street from Tiffany & Co., located at 727 Fifth Avenue in the building that had once been the Huntington family's grand five-story mansion. Jewelers had grown accustomed to waiting out adverse circumstances. But as the Second World War persisted, the kings of the diamond world understood that, despite best-laid plans, it might take more than the thrill of the Jonker or even a splendid House of Jewels to keep their industry from eventually becoming antiquated. The wealthiest citizens might always buy baubles, but who could predict, in a world so unkind and sobering, whether those less fortunate would follow suit? After a Depression and two devastating wars, women bedecked in diamonds might look terribly out of touch; perhaps the new trend would be to disavow them entirely. It was in the industry's best interests to make sure jewels always looked hopeful and never fogyish or, worse, distasteful.

Luckily, someone at De Beers—the real acting proprietor of the diamond world—had an idea.

6

The Sellers

New York and Philadelphia

De Beers was powerful, but not infallible. The company survived World War I and the Depression with the help of Cecil Rhodes's unlikely successor: a middle-class German-born Jew named Ernest Oppenheimer, who fought his way to the chairman's seat with a sense of purpose that would have made Rhodes proud. Like his predecessor, Oppenheimer maintained the utmost fealty to the British crown, and in 1921, the born outsider achieved consummate insider status when George V knighted him for his wartime efforts on behalf of his adopted country. However, there had been compromises along the way. During the First World War, before Oppenheimer's official takeover, De Beers had responded to the decreased global demand for diamonds by lowering prices, even though that tactic ran against everything Cecil Rhodes had built. Also, community relations in South Africa were worse than ever. In 1913, the government passed the Native Lands Act, an addendum to the Glen Grey Act. This law prohibited the majority black population from owning land in all but a sliver of the country. Overnight, family-run farms became illegal, and huge swaths of the population were relocated with no means to support themselves. Their only option was to work in the mines, where the pay was piddling, conditions were dire, and safety was never guaranteed.

De Beers might have dominated the workforce, but the company's leaders were still nagged by their one fundamental weakness: the syndicate could control diamond supply but not demand. This, of course, is the Achilles' heel of commerce. But what if it didn't have to be? In 1938, Harry Oppenheimer—Ernest's intellectual twenty-nine-year-old son, who had only recently joined the family business—traveled to New York to see what could be done about encouraging its American customers. Friends at the "House of Morgan," as J. P. Morgan's finance company was colloquially known, suggested that Oppenheimer meet with another young businessman who might be able to help. At just twenty-one, Gerold Lauck was a Yale graduate and newly minted advertising executive hungry to make a name for himself.

Lauck looked like a young Winston Churchill and worked at the Philadelphia headquarters of N. W. Ayer & Son, which had a Manhattan branch as well. He was a serious man, with a commanding presence and a slight air of melancholia. At that first meeting, Oppenheimer asked Lauck to put together a proposal for De Beers, outlining the way that advertising could make a difference with diamond sales. He assured his new acquaintance that, should De

Beers hire N. W. Ayer, it would be its exclusive representative in America—the market it was focusing all of its energies on, given the problems in Europe. More, De Beers would pay for any research Lauck conducted in order to better understand prevailing attitudes toward gemstones. Gerold Lauck, who was looking to add another marquee client to a list that already included Goodyear Tires and Hanes Knitwear, and to grow N. W. Ayer & Son after some corporate reshuffling, couldn't think of any reason to say no.

There was, however, a significant challenge inherent in the project. What Oppenheimer suggested—whether he knew it or not—was an entirely new kind of advertising, in which brand recognition wasn't the primary goal. The campaign wouldn't be about elevating a name, like Sears, Roebuck or Tiffany & Co. Given the provisions of the Sherman Antitrust Act and the syndicate's monolithic structure, De Beers couldn't sell directly to American customers even if it wanted to. Yes, Lauck understood that the path to selling diamonds had to be a bit circuitous. Promote the *idea* of gemstones and everybody profited. What was good for Harry Winston was, undoubtedly, good for the people at De Beers.

Years later, long after Gerold Lauck flew to Johannesburg and presented his findings—after he and Harry Oppenheimer made a handshake deal and De Beers signed on as a client—the diamond company and the advertising agency became so enmeshed, and had profited so considerably from N. W. Ayer & Son's work, that the lawfulness of their relationship was questioned by the US government. But not yet: at that second meeting, Lauck warned the Oppenheimers—in no uncertain terms—that the diamond was in trouble and, worse, it wasn't just on account of the Depression. For years, there had been a downward trend in diamond sales, in part due to the economy but also because the younger generation regarded diamond jewelry as passé. Ayer's surveys and interviews implied that even the diamond engagement ring, which just twenty years ago had been the cornerstone of every proposal, was in danger of losing its primacy to other gems—pearls were popular—or, worse, to no ring at all. Still, Lauck believed that advertising could help. He had a plan to metaphorically brush the dust off diamonds and make people want them again. De Beers agreed to a modest budget, and the N. W. Ayer team got started.

◇◇◇◇◇◇◇◇◇◇◇◇◇◇◇◇◇◇◇◇◇◇◇◇◇◇◇◇◇◇◇◇◇◇

The obvious next step, consistent with advertisements of the time, would have been to hire beautiful young women and drape them in jewels. But Lauck wasn't interested in taking the road well traveled. After arriving back in Philadelphia from Johannesburg, he called a young designer named Paul Darrow into his office. Darrow, who was just starting out at N. W. Ayer, listened as Lauck described his idea to use fine art, like sophisticated ink drawings or watercolors, in the new De Beers advertisements. The concept was to confer a level of prestige and romance on diamonds—instead of suggesting that they were somehow new, or cutting-edge, or fresh, Lauck wanted to go against the grain and do the opposite, evoking their elegant history. Lauck recognized that, in many cases, the jewelers themselves had undermined the diamond's cachet by advertising jewels the same way one would sell a lawn mower: with price reductions, fire sales. But a diamond wasn't a lawn mower; it was a topnotch luxury item that consumers should aspire to own. A diamond is a timeless treasure. Could the art department work with that?

There was only one answer implied: yes. For Darrow, the meeting with Lauck was "the most important event of [his] Ayer career."

It was an opportunity to distinguish himself. He, along with Ayer's art director Charles T. Coiner and the agency's copywriters, got to work immediately. Armed with the company's market research as well as Lauck's instructions, they came up with four objectives for the ads. One: "To protect and encourage the engagement ring tradition." Two: "Restore diamonds and diamond jewelry to a position of importance in the world of fashion." Three: "Increase sales of all types of diamond jewelry in the general luxury market in competition with other luxuries." Four: "Educate the public on diamond sizes and prices."

It was no easy task. This last goal—to educate the public—had to do with research showing that men typically shopped for engagement rings alone, as opposed to before the war, when couples were more likely to choose a ring together. This had created a new set of problems. Men admitted that once they got themselves to a jewelry store, they had no idea what to do next. How much were diamonds supposed to cost? What should they expect to spend? There had to be some way for the ads to address this problem. The New York branch, where N. W. Ayer & Son's publicity department resided, was already plotting its own concurrent line of attack.

The first ad ran in 1939, in high-end publications like *The New Yorker*. It was full page, and arresting on account of its bold simplicity. In order to save money, the art department had settled on a two-color palette: the background was plain white, with text and drawings done in black, highlighted by a pale cornflower shade that would come to be known, at least internally at Ayer, as diamond blue. At the top of a frame was an elegant sketch, depicting a memorable moment: a young man and woman look out toward an open, hilly landscape, free from man-made structures except for a distant church spire. Beneath the picture, a quote from the poet

J. T. Trowbridge: "For him the diamond dawns are set in rings of beauty . . ." Then, below that, Ayer's lofty copy: "A young man just engaged is apt to share subconsciously the poet's state of mind. He perceives a world of unsuspected beauties—a future in which the Golden Age is reborn in one predestined couple. Unfortunately for lovers, such a mood, while excellent for the discovery of diamonds in the sky, does not always lead to comparable success on earth. There are many things a man must consider when undertaking one of his lifetime's most important purchases—his diamond engagement ring."

The story goes on to advise on the procurement of the symbol of "a new dynasty," which the young man should expect to vary in price with the differences in "weight, color, quality and cutting." The handy 4Cs of diamonds—color, clarity, cut, and carat weight—were not yet in use, so Ayer's team, who had clearly set their sights on the insecure young lover, had to think of other ways to convey those important distinctions. To that end, the art department created a visual box at the bottom of the page to mirror the sketch at the top. Titled "Current Prices of Quality Diamonds," it depicted four sets of brilliants ranging from one-half carat to three carats, each shown from two angles to exhibit the flat table, and then the pavilion and pointy culet, drawn more or less to scale. It included a price guide: $100 for the lower-quality half-carat option, up to $1,750 for a high-end two-carat diamond. Three carats, at least in Ayer's view, should start at $1,500. This must have been eye-opening to the typical engagement ring buyer who—if he bought a diamond at all—shelled out, on average, less than $80 for the jewel that would ultimately beget his future.

This ad, and other similar ones that ran the same year, contained a line that sounded like advice but functioned as a direct admonishment to the groom-to-be who considered skimping: "The beautiful flame of a diamond is unquenchable. Once you have chosen yours, it will become a permanent symbol throughout your years and far beyond them. However modest, your wife will never relinquish it to more affluent circumstances." The threat is implied: do without a ring, and your wife might leave you. More, a diamond is like an insurance policy: the bigger it is, the better you'll sleep at night.

The campaign, which also promoted romantic gifts of diamonds around Christmastime, functioned as a not-so-subtle push to get men into jewelry stores. Once they were there, however, Ayer and De Beers still had to trust them not to balk, and local jewelers to carry through with the sale. And that's where the New York office came in—or, more specifically, Dorothy Dignam, a.k.a. "Miss Dig." Every morning, the plucky, workaholic publicist showed up at her office on the Avenue of the Americas, wearing her best hat—her signature item—and single-mindedly devoted to the agenda of selling diamonds. From an early age, Dignam felt she was being groomed for a career as a writer. The daughter of a pioneering Chicago ad man, she got her start penning typical women's stories, covering the "Four F's"—food, family, furnishings, and fashion— for the *Chicago Herald* before getting a job in 1929 as the lone female copywriter at N. W. Ayer & Son's Philadelphia location. She was there ten years, during which time she began traveling the world, developing her eye for fashion. At the age of forty-three, the sophisticated Dignam moved to New York (along with her aging widowed mother) to fill a position with the company's publicity team.

Dignam had an outgoing, bubbly personality and an ability

to take charge without making her colleagues feel belittled. She never married, but maintained an active life outside of work, volunteering with organizations that promoted the agenda of women in advertising, and teaching Rockefeller Center's air raid defense class during the war. But she felt most at home at work, and found an opportunity to shine when she was assigned to the De Beers account. While the Philadelphia crew reached for words and images to overtly burnish the diamond's reputation, Dignam's toolbox was more expansive. Among her daily tasks was writing to high-profile women to ask for their best engagement ring stories so that she could in turn share them with the public. She penned press releases to magazines and newspapers, informing her Four F contacts that diamonds were back in fashion. She spent long days at the office, followed by late nights at the public library researching diamond history ("That's New York!" she wrote cheerily to a friend back in Philadelphia). And as for the befuddled young man, with money in his pocket and love in his heart—what would become of him? Dignam put together two sets of pamphlets, for an audience of jewelers on one hand and for the eyes of the groom on the other. The former were pitched as informational tools to help struggling salesmen sell more—and bigger—diamonds. The latter, with titles like "The Day You Buy a Diamond," were written with the expectation that they would be handed out by retailers to their customers as an additional source of friendly and accessible guidance.

Dignam's efforts eventually led N. W. Ayer & Son to formalize two publicity arms of the company, funded by De Beers: the Diamond Promotion Service and the Diamond Information Center. The advertising seemed to work. In the years between 1939, when the first campaign started, and 1941, diamond sales rose 55 percent.

◇◇◇◇◇◇◇◇◇◇◇◇◇◇◇◇◇◇◇◇◇◇◇◇◇◇◇◇

Ayer was doing its job, but there were other possible explanations, external to the diamond industry, for why more men might have been buying rings. When America joined the war effort in 1941, it had a surprising impact on domestic marriage statistics—they soared. The uncertainty of young men going off to the front activated a complementary reflex to honor optimism, to embrace the moment, and to celebrate love as a way to give shape to the unknown future. The war also stimulated the long-depressed US economy, and couples embraced the opportunity to marry after years of putting it off for financial reasons. A 1943 *Vogue* article breathlessly profiled the new trend of "accelerated weddings": "Girls have stopped waiting for their fiancés to finish their internships. They don't wait for Commencement. They don't wait. The war isn't waiting. They take crowded cross-country trains. They marry during furloughs. They marry in a week . . . in a few days. No bride wants . . . or needs . . . to sacrifice any of the beautiful gestures which made a Wedding. She isn't being swooningly, selfishly sentimental. She is being a level-eyed realist." As for the ring, it acted as a perfect and timely visual placeholder for those couples who got engaged, or married, between the man's tours of service. In the same article, *Vogue* acknowledged that "the big, *round*, faceted diamond is again a favorite pledge-thee-my-troth ring"; however, "there are a whole coterie choosing rubies . . . feeling that they somehow symbolize true love." In a separate article from 1942, *Vogue* recommended yet another jewel: "First comes the engagement ring of gold, set with small, natural Oriental pearls and a big cultured pearl in the center."

Pearls and rubies were lovely, but they weren't De Beers diamonds, and this meant that for N. W. Ayer & Son there was still more work to do. A publicity memo written to the attention of Vyvyan Donner, the fashion editor at Fox Movietone News, a sound newsreel—dated November 11, 1940, and entitled "Diamonds and the Draft"—showed just how conscious the people at Ayer were of the way external forces affected other social changes: "A wedding ring manufacturer announces that his wedding ring sales are already 250% over last year. The second largest ring manufacturer in New York says that his business, at present, is 500% ahead of last year—with all 12 of his salesmen on the road selling matched sets of diamond engagement rings and wedding bands. . . . Despite reports to the contrary, this is *not* all due to the draft. Like the Mounty, the American girl always gets her man."

She always gets her man—and her jewelry. Thanks at least in part to Dorothy Dignam's efforts, the media started reporting on the revival of the engagement ring trend—or, in some cases, ignoring that it had ever dwindled. Some articles, like one called "Diamonds Remain Symbol of Man's Ideal of Romance," published in the *Atlanta Constitution* in March 1940, read overtly like De Beers ad copy: "The earth may heave. Stars may fall. The moon may wane. Fashions may die and customs shift. But one thing remains fixed, static, forever unchanged. And that is man's ideal of romance. And the symbol of man's ideal is the diamond." Others, like February 1941's "Solitaire Returns to Favor for Engagement Rings" in the *Washington Post*, were more service-oriented and measured: "The solitaire, or one fine diamond in a plain setting, is back in fashion for engagement rings. And this type of ring represents the best value for your money, because you're not paying for the labor required to cut a lot of small stones." Even reigning empress of etiquette Emily Post's 1940 response to a single young woman who wished to wear an inherited solitaire on her left hand acknowledged the symbolism associated with a diamond: "Although wearing a ring on the left hand does not announce an engagement, a solitaire would naturally suggest [one]." Yet her answer to another reader who inquired whether a platinum diamond ring is the "thing" in engagements likely left the team at Ayer feeling the fight was yet to be won. "It is the right of the bride to choose what she likes best!" Post advised, thus giving the young woman permission to skip the diamond if there was something else she'd prefer. In one internal memo, George D. Skinner—Dignam's superior at the New York office—remarked that, unfortunately, Emily Post couldn't be bought.

Meaning the female consumers themselves would have to be convinced.

<hr />

Around the same time that the market for gem-grade stones was floundering, sales of industrial diamonds—used in everything from drill bits to abrasives to car and airplane parts—were booming. Demand for bort, as the industry nicknamed by-product stones that had no future as jewels, skyrocketed as the war waged on and national defense budgets expanded. When the United States joined the war effort, tensions between the Roosevelt administration and De Beers rose until the American government explicitly accused the syndicate of hoarding the industrial diamonds that were needed in factories and thus critical to the Allied war effort. Ever the capitalist enterprise, De Beers swore its loyalty to the cause but refused to release any measure of diamonds—bort or others—that might threaten its stockpile. In retaliation, the US Justice Department opened an investigation of De Beers in order to, among other things, determine whether or not it was using N. W. Ayer & Son as a cover to bypass the Sherman Act, which prohibited monopolies or cartels from operating on US soil. The accusations against Ayer, as a delegate for De Beers, mounted, and by the time the war was over, they ranged from the intangible—deceiving the American public into thinking diamonds were rare—to the more explicitly criminal: acting as liaison in underground diamond sales.

As De Beers tussled with the US government over industrial diamonds, N. W. Ayer & Son continued releasing new ads encouraging women to want diamonds and men to buy them. A series that ran in *Vogue* through the early 1940s emphasized the heirloom and emotionally autobiographical nature of diamonds: "You will find more than starry gems and ornaments pillowed on the satin of her jewel case," one 1941 page read, under a framed portrait of a gentlewoman that looked as if it belonged on a museum wall. "There you will find her memories." Another 1942 ad evoked the war without explicitly naming it: "Time and circumstances and thousands of miles have lain between them more than once in their eventful years of marriage . . . And in his gifts—those later, more lavish diamond tokens of his homing love—she cherishes the joy of each reunion crystallized in brilliant reassuring flame."

The *Vogue* ads emphasized the idea of a diamond collection, one that begins with the solitaire and grows with each milestone event—as does, ideally, the couple's love. This was Ayer exploring the premise that the engagement ring didn't have to be just a once-in-a-lifetime gift; rather, in the best circumstances, it could function as a gateway diamond, fusing in the bride's mind the connection between joyful occasions and gemstones and creating a socially appropriate context for her to start feeling comfortable wearing them. As for the importance of the engagement ring itself, *Vogue* readers were getting the message from all sides. Recognizing that weddings and engagements were fashionable, the Pond's Extract Company, along with its advertising agency, J. Walter Thompson, launched a campaign for its most famous product—Pond's Cold Cream—using close-up beauty shots of neatly groomed socialites with the tagline: "She's engaged! She's lovely! She uses Pond's!" The ads relayed their models' engagement stories and even featured photos and descriptions of the ring. As the war wore on, the features got more patriotic in nature, introducing young women whose fiancés were overseas, and who were doing their part by working in factories. Still, their diamond rings were invariably showcased, and the copy always ended on the same chirpy note: "It's no accident so many lovely, engaged girls use Pond's!"

These ads positioned an engagement as a terrific accomplishment for a young woman; of course, success was imminent with the right skin cream. And as for the diamond ring, sparkling on the "heart finger"—it was her badge of honor, her trophy. Who would complain if her prize shone a little brighter than everyone else's? Soon, men and women alike would have access to a more specific, universal vocabulary to discuss and compare diamonds, beyond just their own layman impressions. Thanks to De Beers, N. W. Ayer & Son, and a relatively new group called the Gemological Institute of America (GIA), Americans were about to become educated in gemstones. After all, if a woman could tell the difference between a lower-quality diamond and a more impressive one, wouldn't she yearn for the latter? More important, if a man knew that there was an amateur jeweler waiting for him at home—and, truth be told, if he understood the finer points of diamond quality, too—wouldn't his inclination be to spend more money on a ring?

∞∞∞∞∞∞∞∞∞∞∞∞∞∞∞∞∞∞∞∞∞∞∞∞∞∞∞

The GIA was a Los Angeles–based organization devoted to the cause of training American gemologists. It was a family-run business created by a man named Robert M. Shipley, a midwestern jeweler who realized the hard way that he really knew nothing about his merchandise. One day, an oil baron from Tulsa, Oklahoma, came into Shipley's store. Frank Phillips wanted to buy a truly magnificent bauble for his wife and suggested an emerald ring. Shipley ordered a hundred thousand dollars' worth of the finest emeralds for the client to choose from. As the pair sat in Phillips's gilded parlor, the client asked how to tell the difference in quality between the gemstones on offer. Shipley was put on the spot. He fumbled and then parroted some trivia he'd once read in a book on mineralogy. But Phillips could see right through him: Shipley didn't know what he was talking about. The customer's confidence was damaged, and he took his business elsewhere.

Shipley was embarrassed, but he came to understand that there were many others out there like him, who were unwittingly devaluing precious stones by shilling for a product without having developed any kind of expertise. He studied in Europe, and then dedicated the rest of his life to the mission of educating American jewelers about gemstones.

De Beers, N. W. Ayer & Son, and GIA all like to take credit for introducing the concept of the "4Cs," a memorable shorthand for the four aspects of diamond grading: cut, color, clarity, and carat weight. While it's unclear which team actually started using the term first, it is apparent that during and after the war, the diamond industry made a concerted, synchronized effort to standardize its vocabulary, with the hope that a common language would convey a level of authority and prestige to sellers, while simultaneously em- powering buyers to feel more comfortable talking about diamonds.

But none of those C-words were as influential as a phrase that showed up in the ads a few years later, in 1948. One afternoon Gerold Lauck called Frances Gerety, a young copywriter, into his office. On his desk there was a mock-up for the next round of print ads, which evoked the same aristocratic character, the same dreamy elegance, as earlier entries in the campaign. However, Lauck was unsatisfied. Although the ads still featured hand-drawn pictures of diamonds, he wanted something else at the bottom of the page to tie the whole thing together—something that immediately identified it as dia- mond advertising. Gerety, who had taken over the copywriting for De Beers after her predecessor retired, wasn't terribly daunted by the prospect of coming up with a tagline. Other people were afraid of Gerold Lauck, but she was not. For Gerety, he was a brilliant man, something of a role model, with, yes, an unpredictable streak, too— but what was so bad about that?

Gerety had started working at Ayer four years earlier when, at the age of twenty-eight, she called the company looking for a writ- ing job and somehow found herself talking to a bigwig who re- marked that her timing was perfect: they had just lost one of three lone females on the sixteen-member copy team. Like many adver- tising firms at that time, Ayer hired women to handle the "softer" accounts—soap, toothpaste, hosiery. Diamonds, because they were linked to love and romance, slid easily into that category. Gerety quickly proved her talent. She had the irresistible combination of a woman's perspective and a sharp, unfussy ear for language. She was precise; she never would have tangled with anything like the tagline that a previous writer had suggested: "Your Diamond is a permanent treasure, and you buy it—not for today—but for all your yesterdays to be." It was convoluted and, unsurprisingly, had never made it into print.

Frances Gerety was also a terrific procrastinator, and sure enough, by the time the copy was edited and the art department had done its job, she still hadn't written her pièce de résistance: that one encapsulating thought. A self-described night owl, she had taken the layouts home the evening before they had to be packed up and ready for Gerold Lauck to take them the next morning to Johannesburg, where the Oppenheimers would give their approval. At four a.m., the papers were sitting on Gerety's nightstand and she was still mentally tinkering.

"Dear God, give me a line," she thought. She wrote something down and, more exhausted than satisfied, drifted into expectant sleep.

The next morning, she reread it: "A Diamond Is Forever." Not exactly groundbreaking, but interesting enough, and Gerety hoped that her superiors at N. W. Ayer & Son would agree. She showed it to Lauck and the copy chief, George Cecil; as she recalled later, "Nobody was jumping up and down about it." Conservative Cecil wondered aloud if it was even proper English. Gerety found herself arguing in favor of the line that had barely made an impression on her a few hours earlier: "As far as I'm concerned, the word 'is' means it exists. . . . It's a synonym for exists."

Ultimately, the team decided to go with the line; after all, it was just one piece of the ad, and they didn't have anything better to use. "A Diamond Is Forever" debuted along with a colorful update to the campaign. Who knew that a tagline could be forever, too? As Gerety eventually recalled in a letter to Ayer's vice president: "I shudder to think of what might have happened if a great line had been demanded—every copywriter in the Department coming up with hundreds of lines, and the really great one lost in the shuffle."

"A Diamond Is Forever" marked De Beers ads going forward— ads that became increasingly elaborate as Ayer's budget allowed them to expand from the original two-color palette. Art director Charles T. Coiner had a persistent nature and an eye for talent, and he commissioned work from artists as grand and varied as Pablo Picasso, André Derain, and Salvador Dalí, showing up at the last's door in Paris, where Dalí's wife initially turned him down because the painter was feeling sick after trying an experimental diet that involved eating human hair. The ads, while a bit more eye-catching now, continued to evoke the same aspirational pleasure as the earlier versions. Churches, countrysides, weddings, and honeymoons were all consistent visual motifs.

The print ads were effective, but they were only one arm of the diamond industry's multilimbed postwar campaign. The people at N. W. Ayer were dedicated to the task of building a social custom. For that, they needed to cultivate role models: women who had status and charm, and could make diamonds look great.

7

The Queens

London, South Africa, and New York

There are two kinds of fashion plates: those who are deliberately created and those who emerge, through a series of unexpected circumstances, as objects of fascination the world over. Wallis Simpson was the second type of icon. She was a humble American from Baltimore, twice divorced and well into her thirties by the time she met and charmed His Royal Highness the Prince of Wales, next in line to the British throne.

Bessie Wallis Warfield was born into a prominent Maryland family, but her mother fell upon hard times after her father passed away. They relied on financial support from relatives throughout much of Wallis's childhood. When she was old enough, she married a US naval officer whose position with the armed forces conferred a certain level of respectability and prestige on the couple. However, the uniform was a smoke screen; Win Spencer was no gentleman. He was an angry drunk who abused his young wife at whim, and though Wallis tried to endure it, she ultimately had to save herself and escape the marriage.

Wallis moved to Britain with her second husband, Ernest Simpson, a smartly dressed, Harvard-educated shipping executive whose business was abroad. She reinvented herself in London. Growing up alongside Washington, DC, society, she had absorbed the taste and manners of her peers, even if she didn't necessarily have access to the same resources. In London, she found herself rubbing elbows with a highbred group of expats and English aristocrats, who quickly accepted the elegant American couple into their social scene. Wallis Simpson was clever and flirtatious, so she was an appealing guest, and she loved meeting interesting people even if deep down she suspected she didn't exactly fit in. Not quite a full-fledged beauty, Simpson was serpent thin with striking features: she had a prominent nose and an almost masculine jawline, balanced by a silent film star's soulful eyes and soft mouth. To her frustration, money was still an issue: although Ernest Simpson's business had done very well historically, the wars in Europe and the Depression in America cut a huge slice into his profits. That didn't stop Wallis from going to dinner parties and hosting her own "KTs"—cocktail hours—at their home. She was a stylish dresser who could create the illusion of an enviable wardrobe from just a handful of fashionable pieces. She also had a benefactor in her aunt Bessie, who seemed happy to send regular checks from America with the expectation only of receiving a steady stream of letters from her niece.

Simpson first met the sweet and delicate-featured Edward, Prince of Wales, through her close friend Thelma Furness, who was publicly dating him. The daughter of a diplomat and a member of American society, Furness thought little of running around town with a prince, but Wallis was more impressed by his status, confessing to Aunt Bessie, "I've made up my mind to meet him." She accomplished this easily, introducing herself at an intimate party of just seven guests. But it wasn't until a few months later that Simpson made any kind of lasting impression on the man who would eventually become absolutely smitten with her. Every year in the late spring, at the dawn of a new social season, a number of women were invited for a presentation at court, where they dressed up and curtsied for the royals. The purpose of this was largely so that ripe, marriageable socialites made their debut in full view of suitable bachelors. But on occasion, older married women were also included, provided they had the right connections. In Simpson's case, her "debut" in 1931 at the age of thirty-four was symbolic, signaling her and her husband's official entrance into the world of the British upper crust.

Her newfound status notwithstanding, she still found herself in a borrowed white satin gown, train, and feathered headpiece: the traditional garb of an English debutante. Simpson splurged on her own jewelry, likely using Aunt Bessie's money to buy a long aquamarine-and-crystal necklace—not diamonds, as they were too costly—with a large dangling pendant in the shape of a cross. Escorted by her husband, Simpson felt flutters in her stomach as they waited their turn to meet the sovereigns, King George V and Queen Mary. Later, as the prince strolled by her, she heard him whisper to his uncle: "Something ought to be done about the lights. They make all the women look ghastly."

Overhearing this unchecked comment from the prince delighted Simpson, who found the pageantry of the day both awe-inspiring and a little bit silly. That night, she attended Thelma Furness's celebration, where "HRH"—His Royal Highness, as Simpson had begun referring to him in her letters to her aunt—made a late-night appearance. When he complimented Wallis on her dress, she saw an opportunity. "But, sir," she said, "I understood that you thought we all looked ghastly."

The prince was taken aback—after all, just who was this brash American?—and then smiled broadly, confessing he didn't know that his voice would carry. But for Simpson, it didn't matter if she'd momentarily embarrassed him. Finally, she'd done something to get the prince's attention. She wasn't angling for his affections, exactly—she had Ernest, and he was Thelma's beau—but she was determined to become a part of his inner circle. As the evening came to a close, Prince Edward offered the Simpsons a ride home in his car, where he talked about making improvements to his country home, Fort Belvedere, and spending long mornings at work in the garden. This kind of small talk might have bored Wallis in a different context, but given that the narrator was the eventual British monarch, she hung on his every word.

When the prince dropped the Simpsons off at home, they made polite overtures to see one another again. It didn't happen right away, but the couple, and especially Wallis, became fixtures in HRH's social life. They spent long summer weekends at the Fort, and Wallis and the prince struck up a friendly correspondence. By early 1934 the two were so close, sharing everything from decorating tips to heartfelt advice, that when Thelma Furness planned a trip back to New York for the winter, she asked Simpson to "look after" the prince for her.

She had a glimmer in her eye as she said it. Perhaps she had grown tired of her high-maintenance companion—he was desperate not only for affection but also for a constant supply of almost parental guidance—and was willingly passing him off to an eager, albeit married, surrogate. But even if Thelma Furness had that kind of foresight, she never could have guessed what would happen next.

∞∞∞∞∞∞∞∞∞∞∞∞∞∞∞∞∞∞∞∞∞∞∞∞

Wallis Simpson was not the only foreigner who was mesmerized by the British royal family. Far from it: Americans in particular found the Windsors enchanting, starstruck by the very sense of ceremony and entrenched power that their forefathers had fought against more than a century and a half ago. In 1947, years after Simpson had effectively joined the ranks of British high society and after the Second World War was over, De Beers sought to capitalize on the royal family's overwhelming popularity. The Windsors had been committed diamond lovers and collectors throughout their reign—which was good for the company's profile—and had recently impressed the world with their decision to stay in England for the duration of the fighting instead of fleeing to safer shores. At the earliest opportunity, De Beers invited them down to South Africa for a tour of the mines, most likely knowing it would create an irresistible press moment. The trip was good for the crown, too; although South Africa was no longer a British colony, the Central Selling Organization—the subsidiary of De Beers that handled rough diamond sales—was based in London. On the afternoon of January 31, King George VI—Prince Edward's younger brother—and Queen Elizabeth, along with their daughters, Princesses Elizabeth and Margaret Rose, boarded the battleship *Vanguard* while over five thousand English citizens braced against the freezing winter weather to watch. They set sail early the next morning flanked by an escort of destroyers, leaving the pockmarked town of Portsmouth, still scarred by Nazi attacks, in the distance.

The royal family arrived at Cape Town almost three weeks later, disembarking into 113-degree (F) heat. There, too, they were greeted by hordes of admirers; as the *New York Times* noted: "Capetown's interpretation of the color bar is the most liberal in this country; in the crowds that lined famed Adderley Street and other main thoroughfares descendants of British and Dutch settlers stood elbow to elbow with Indians, Malays, Negroes and persons of mixed blood, and all cheered in unison. . . . Those who had stood patiently for hours on the city's sidewalks were not perplexed by the fine points of etiquette—they roared their greetings, each in their own tongue." The Windsors planned to stay for three months. Amid speculation that the king and queen had political or imperialist motives for their trip, talk in the media also turned to diamonds even before the royals set foot on the dry Cape Town soil. When increasingly critical reports started surfacing that Princess Elizabeth would receive four hundred diamonds valued at $100,000 (or more) for her twenty-first birthday in April, which she would celebrate in South Africa, the government responded by releasing an official correction stating that she would be given just twenty-one brilliants, while Margaret would receive seventeen cut and polished stones of her own.

Apparently, that number didn't include the fifteen thousand dollars' ($163,000 today) worth of diamonds presented to Elizabeth by South African Railways and Harbors at a ceremony on March 4, when the charming princess attended the inauguration of a dry dock named for her. Nor did the estimate reflect the six-carat rock, worth about $6,000 ($65,000 now) she was offered two days before her birthday celebration by the three-year-old Mary Oppenheimer, granddaughter of Sir Ernest. The *Chicago Tribune* reported that the young girl curtsied shyly as Elizabeth graciously exclaimed, "What a wonderful present, I am delighted!" and the elder Oppenheimer assured her it was a "perfect blue white stone." The same article reported that the princess would indeed be given twenty-one "major" stones on her birthday—the largest at 9.55 carats—totaling 71.31 carats, but she'd also get an additional sixty-six smaller diamonds with which to complete a stunning necklace. Not to be slighted, Princess Margaret received her own 4½-carat stone from Mary Oppenheimer—valued at $4,000 ($43,000 now)—and fifty-three diamonds weighing 41.57 carats, which she might use to make a bracelet. Their mother, the queen, was said to have accepted an 8.55-carat marquise solitaire, while King George was himself the recipient of 399 diamonds, arranged and presented in the outline of a springbok: a lithe antelope native to South Africa.

Whether or not these reports were exaggerated, they had the effect of keeping diamonds in the headlines and building an association between gemstones and the well-liked elder princess, who one day soon would succeed her father as Britain's highly influential sovereign. For Princess Elizabeth's birthday, the Oppenheimers treated her to a tour of the now dried-up Kimberley mine, whose central crater, once nicknamed the Big Hole, had been left untouched as a tourist attraction. At the Kimberley Club afterward she met the industry leaders, who no doubt vied for a moment or two in her presence. The princess's role as a jewelry icon was not confined to the shores of South Africa; once the royal family returned home to London, they launched into preparations for Elizabeth's marriage that same November to Lieutenant Philip Mountbatten, soon to be created the Duke of Edinburgh. Come September, the papers were filled with speculation and a slow but constant trickle of details about the upcoming Westminster Abbey wedding, from descriptions of the four-tier cake to rumors that landlords who were lucky enough to own space along the procession route were charging upwards of $140 ($1,500 now) for an unobstructed view. As for the wedding gifts, they were put on display for weeks before the wedding at St. James's Palace, where the public was invited to ogle them. The jewels were especially striking: a ruby-and-diamond necklace from the king and queen; a diamond tiara, brooch, and stomacher from the bride's grandmother Queen Mary. The 1,200 people who had been so kind as to send gifts were invited to a private reception, where those who came in contact with the soon-to-be bride might have noticed her engagement ring: a platinum-and-diamond piece with a three-carat round center stone and five diamonds on either side, all repurposed from one of Philip's mother's own tiaras.

At the November 21 wedding, the princess wore the "fringe" tiara made by Garrard in 1919 for her grandmother Queen Mary, who had given it to her daughter-in-law Elizabeth, the bride's mother. Queen Elizabeth lent the piece to her daughter on her wedding day as her "something borrowed," and the princess walked down the aisle in the spiky all-white diamond crown made of alternating jeweled pickets, half finished with pointed edges and the other half rounded. She also wore two strands of pearls given to her by her parents, as well as a pair of pearl-and-diamond earrings—the center pearls each surrounded by eight brilliants—that had been made at the turn of the century and had once belonged to her grandmother.

The wedding was covered extensively by the American press,

featured in magazines like *Vogue*, *Life*, and *Reader's Digest*. Overall, the festivities were lovely and less lavish than they could have been, by the standard of other royal weddings, given the proximity to World War II. But that didn't stop young women from wanting to imitate the bride—if anything, the restraint only made her feel more accessible. As Penny Proddow and Marion Fasel put it in their book *Diamonds: A Century of Spectacular Jewels*: "Enchanting as a fairy tale, [Princess Elizabeth's] wedding had the side effect of making formality fashionable and appropriate. The images of Elizabeth and the nobility wearing diamond jewels made a deep impression."

◇◇◇◇◇◇◇◇◇◇◇◇◇◇◇◇◇◇◇◇◇◇◇◇◇◇◇◇◇◇◇◇◇◇◇

Incredibly, none of that—not the family's diamond-studded trip to South Africa, nor the tastefully glamorous royal wedding—would have been possible if it weren't for Wallis Simpson. Without her, there would have been a different king. After Thelma Furness left London, Simpson kept her word to "look after" the prince, eventually having him over for dinner many nights a week. He was needy, certainly, asking her to mind his social calendar and staying into the night after each meal so that Simpson soon found herself exhausted. However, she couldn't ignore his sweetness and generosity. He showered her with expensive jewelry from Cartier, Van Cleef & Arpels, and Harry Winston, as well as with clothing and other opulent presents, and her collection of high-end baubles soon became the talk of the London social circuit. Simpson's days of saving up for an aquamarine necklace were behind her. For her part, she couldn't quite get over the surprising fact that the prince seemed genuinely taken with her. It was just so flattering, especially considering how Simpson saw herself. "I certainly was no beauty," she wrote in her memoirs about the early days of the affair, "and he had the pick of the beautiful women of the world. I was certainly no longer very young. In fact, in my own country I would have been considered securely on the shelf."

She was also married, though the relationship with Ernest had been strained by their money troubles, and further damaged by Wallis's ongoing dalliance with HRH. Whether or not she was technically cheating—that point is still debated—she made a priority of her regal companion, and in doing so treated her husband as second best. Then, at the end of 1935, when Wallis was almost forty, the sickly king passed away and David—as the Windsors referred to Prince Edward and as Simpson had taken to affectionately calling him—succeeded him. Suddenly, Wallis Simpson was engaged in what appeared to be an adulterous affair with the king of England, and the American and European gossip columns started paying attention. (The British papers, in deference to the royal family, ignored her.) David made it clear that he wanted to marry her, which was both pleasing and frightening. He seemed indifferent to the repercussions, even though in order for marriage to be possible she would have to divorce Ernest in a Christian realm, where no one who mattered—not the Windsors, nor Parliament, nor the Church of England—approved of the king bedding down with an immoral American.

Wallis became "Wally" in the tabloids: a vicious and money-grubbing character who would stop at nothing to get her claws in the otherwise popular young king. It's true that David gave Wallis so much money that she implied to Aunt Bessie that she'd never have to worry about her finances again. But she also had second thoughts about the relationship, especially when David implied that he would stop at nothing until she was his bride. She tried to convince him she was happy with Ernest, who by then had his own lover. "True we are poor and unable to do the attractive amusing things in life which I must confess I do love and enjoy," she wrote to HRH. "Also the possession of beautiful things is thrilling to me and much appreciated but weighed against a calm congenial life I choose the latter." Still, he was undeterred. While the gossip columns reported on her alleged manipulations, the truth was that the king had finally found someone to love and even the reality of a jilted husband wasn't enough to give him pause. Some speculate that King Edward approached Ernest himself after Wallis tried to break things off—her letters reflect a change of heart—and used his clout to convince Ernest to leave her.

On December 11, 1936, Edward VIII shocked the world by announcing that he had chosen to abdicate the throne rather than give up his controversial American beloved. He shared the news over the radio, and Simpson listened from a friend's hideaway in Cannes with her hands over her eyes. She was already ostracized, and she knew she'd be blamed for nothing less than disrupting Great Britain's entire political landscape. Even so, part of her appreciated that David's choice was the ultimate grand gesture—the twist ending in a captivating reverse (and, to some observers, perverse) fairy tale. In Simpson's story, the unconventionally attractive woman from a modest background in Baltimore didn't end up living in a palace like a traditional heroine. Rather, she proved so spellbinding that she caused a king to do the unthinkable: walk away from his rightful crown.

And what crown would that be, exactly? As soon as King Edward VIII made his shocking announcement, his subjects started wondering about jewelry. The very same day of the abdication the *Chi-*

cago Tribune published an article describing the actual crowns that he and Simpson would no longer have at their fingertips now that his eviction from Buckingham Palace was ensured. The *Daily Boston Globe* ran a piece about Simpson, reporting on the "bitter question which secretly agitated high British circles": "Would Mrs. Simpson 'get away' with royal jewels?" Apparently, the king had lent her baubles from the royal coffers, and some vocal critics worried that the temptress would abscond with them. But in truth, Wallis Simpson had plenty of treasures of her own, which the king had been showering her with since his days as a love-struck prince.

During the second half of her life, Wallis Simpson, who became the Duchess of Windsor when she married Edward, was a darling of *Vogue*. Her unusual face was a cipher, and the same question that defined her entry into the newspaper columns—what is it about her that makes a man want to give up his kingdom?—continued to give her global allure. "What is it that puts the Duchess of Windsor at the top of every 'best dressed' list," the *Daily Boston Globe* asked in 1948. "It isn't her face; she isn't pretty. It isn't her figure; her head is slightly larger in proportion to her body." Maybe it had something to do with her accessories. The former king made a practice of not only buying his paramour the most expensive, high-end jewels from the best houses in the world, but also commissioning many pieces exclusively for her. There was the ruby-and-diamond bracelet from Van Cleef, which Edward sent her during the difficult period after his abdication, and which he had inscribed with the date—"27-III-36"—and the words "Hold Tight": something of a mantra for the pair during their forced separation, while they were still waiting for her divorce to go through. Another Van Cleef piece arrived on Simpson's fortieth birthday: this one a stunning diamond collar that looked to be wrapped with an overlay of rubies, its red tail cascading down from the side. It also came with a simple yet romantic inscription, "My Wallis from her David."

Her Cartier pieces were more outrageous. The war had been especially challenging for Cartier, not just because the Place Vendôme was under fire and platinum was strictly sanctioned but also because Louis and Jacques Cartier both died during that time. In 1945, just five years after the second World's Fair, Pierre Cartier said good-bye to his New York store, in the beaux arts mansion on Fifth Avenue, which he had purchased from industrialist Morton F. Plant for the price of $100—plus a double strand of top-quality natural pearls for Plant's wife. He returned home to France to take over Cartier Paris in the absence of his brother. There, he worked alongside artistic director Jeanne Toussaint. The designer had been handpicked by his brother Louis to work for Cartier after he saw her handbags, even though she couldn't draw and had no background in jewelry. It didn't hurt that Toussaint was also Louis Cartier's lover. Toussaint brought an element of unrestrained whimsy to the house, and designed a number of pieces for the Duchess of Windsor. Toussaint, who during the Nazi occupation integrated subtle patriotic imagery into her designs, was a bit of a subversive herself, and Wallis Simpson's biographer Anne Sebba contends that the two women, while not necessarily politically aligned, still understood each other.

Among the most celebrated pieces Toussaint designed for the duchess was a flamingo brooch in 1940; the duke had commissioned it when he made an unexpected trip to Paris on the eve of the war. Toussaint, along with her design partner Peter Lemarchand, loved experimenting with animal shapes, and this piece was stunning for the delicacy of its pointed beak and spindly legs, as well as its bold use of color, with white diamonds for the body and a splashy wave of feathers made of rubies, sapphires, and emeralds. Wallis Simpson wore the brooch throughout her life. She also helped make the panther the house's signature image. Since the early twentieth century Cartier had been experimenting with diamond-and-onyx pavé (French for paving), where the small gems hugged together like black and white cobblestones. Eventually, they nicknamed the design "panther skin" for the way it resembled the wild cat's spotted coat, and it was used for everything from bracelets to watches. The technique ultimately engendered a more literal translation, with Toussaint—nicknamed La Panthère—creating three-dimensional figures of cats.

One of her earliest designs of this type was a brooch from 1949, with a small delicate panther made of diamonds and sapphires perched aloft on a huge 152.35-carat cabochon (smooth polished, without facets) sapphire. The piece was made for stock—meaning without a buyer or wearer in mind—but the duke purchased it for his wife. Simpson must have liked it, for in 1952, Edward commissioned another panther just for her. This one—an articulated bracelet with the classic diamond-and-onyx skin—wrapped around the duchess's wrist in such a way that the animal appeared ready to pounce, with his head perched on the bone, his emerald eyes alight, and his paws extending even beyond the arm.

Simpson had no need for the crown jewels—she had her own collection, growing by the day. As early as 1946, when her English country apartment was burglarized, it was estimated that her collection was valued at over $1 million, the equivalent of about $12.2 million in 2014. In 2010, when Sotheby's London hosted a posthumous sale of the duchess's jewelry, the Toussaint panther commanded a record 4.5 million pounds (over $7 million now), making it the most expensive bracelet ever sold at the time. The sale put the piece back in the news, and as photos surfaced, observers debated whether it was gorgeous or grotesque. A unique work of art, or tacky trompe l'oeil? Fitting, given that the same question has been asked about Wallis Simpson herself.

∞∞∞∞∞∞∞∞∞∞∞∞∞∞∞∞∞∞∞∞∞∞∞∞∞∞

The fact that the Windsors seemed to genuinely appreciate and enjoy diamonds was a boon for the industry—one that De Beers and the high jewelers were more than happy to exploit during the prosperous years following the war. But across the pond, press-savvy Harry Winston, a patriot, wished that America had its own equivalent of a royal family who, by using their celebrity to endorse jewels, made average citizens admire and covet them. The United States had emerged victorious from the war, and Winston felt it was only appropriate that its citizens should stand proud and start embracing their newfound wealth and global status. To convey an overall sense of affluence, Winston started lending his pieces to top-notch brands like Cadillac, Parliament, and Revlon for use in their ads, but like the people at N. W. Ayer & Son, he understood that classic print advertising could accomplish only so much.

By this time, Harry Winston was located at Fifth Avenue and Fifty-First Street, in a stone mansion once owned by a countess, steps from Tiffany & Co., Van Cleef & Arpels, and Cartier. In a 1953 profile of the famed street, the *Washington Post* dubbed Fifth Avenue a "Jeweler's Showcase." Harry Winston—described in the same article as "one of the most unusual enterprises in the glittering half-mile"—intended to bring that showcase to anyone and everyone who, for whatever reason, didn't have an opportunity to amble down the streets of New York. At one point, he lobbied the

White House to put together a national collection of jewels akin to the ones belonging to the European crowns, which he envisioned would be at the disposal of the First Ladies, who had countless international events and soirées to attend. However, the idea was shot down as too aristocratic, and thus contrary to the ideals of a democratic nation.

And so Harry Winston came up with his own personal nod to royalty. The Court of Jewels was a traveling exhibition of spectacular gemstones that Winston had acquired over the course of his career, including the polished Jonker (formerly known as Number One), the Idol's Eye (a pale blue 70.21-carat diamond believed to have been mined in Golconda in the seventeenth century), and the Star of the East—the stone Pierre Cartier once sold to Evalyn Walsh McLean on her honeymoon. Harry Winston purchased the last in April 1949, after the infamous Washington, DC, socialite passed away. Although McLean's will stipulated that her jewels should be held in a trust until October 25, 1967—when her youngest grandchild turned twenty-five—the family petitioned the court for permission to sell the collection earlier in order to pay their debts. There were plenty of potential buyers, but Harry Winston swept in and offered to take the whole lot, for an estimated price of somewhere between $1.25 million and $1.5 million (over $13 million now).

This, of course, made Harry Winston the new owner of a stone even more sensational than the Star of the East: the Hope Diamond, the most notorious gem in America. Like Pierre Cartier, Winston wasn't deterred by the threat of the curse—but he certainly enjoyed the lasting impression the so-called hoodoo diamond made on others. Soon after he acquired it, Winston and his wife, Edna, decided to fly home separately from Lisbon to New York—the idea being that if an accident occurred their two young sons wouldn't be orphaned. Edna Winston left first, and the next day Harry got on his flight cheered by a telegram that had just arrived confirming her safe arrival. Winston found himself seated next to a man who, in an attempt at friendly chatter, confessed that he'd narrowly escaped disaster the day before when he learned he would be flying alongside Mrs. Harry Winston. Immediately, he changed his travel plans to avoid the consequences of sharing an airplane with someone who might have touched that unlucky stone. He wondered aloud if the plane had even landed. Without a word, Harry Winston pulled the telegram from his wife out of his pocket and passed it over—effectively silencing his seatmate for the rest of the trip.

The Hope was the star player in the Court of Jewels lineup. Almost everyone had heard of the diamond, but as it turned out, very few people knew what it actually looked like. The exhibition debuted on November 15, 1949, with a party at the Ritz-Carlton, and then opened down the block at Rockefeller Center, where admission was 50 cents for adults (about $5 now) and 25 cents ($2.50) for children, with proceeds going to the United Hospital Fund. It was designed to convey a sense of splendor; visitors entered by way of a red staircase and passed an imperial fountain on the way to the Hope. Viewers might have been familiar with the tale of the stone—which was lengthily reiterated as part of the display—but they were reportedly shocked by its appearance. "To most people's amazement, the thing turned out to be dark blue instead of white," the *Hartford Courant* recounted. "One woman said it wouldn't do at all with her new purple velvet evening gown and a man with a southern accent said he thought the white ones were shinier."

Good thing it wasn't the only attraction. Audiences at Rockefeller Center were treated to a diamond-cutting demonstration, and as the Court of Jewels made its way around the country over the next five years, Americans jumped at the chance to see twelve million dollars' worth of precious gemstones. Maybe Americans didn't need their own queens and princesses in order to find fine jewelry captivating. Besides, there was a place out West with a different kind of potential from Buckingham Palace. The United States didn't have royalty, but it had something else, possibly something with even stronger commercial sway: it had the glitter and magic of Hollywood.

More specifically: in a dark sky arching over the desert, it had stars.

8

The Stars

Los Angeles

There's no advertising like free advertising, and for the diamond industry, *Gentlemen Prefer Blondes*—the Broadway musical and then the film, based on a best-selling 1925 novel by humorist Anita Loos—was the ultimate gimme. In 1949, comedian Carol Channing took the New York stage as Lorelei Lee: an airhead from Arkansas and the quintessential gold digger with a heart of . . . well, gold. Loos's smart satire hinged on the premise that readers were in on the joke; that they could see Lorelei for who she really was—an unreliable narrator who measured her worth in social cachet and material possessions. By casting Channing, the show's producers bet her presence—her gawky long limbs, raspy voice, laughing eyes, and amusingly large mouth—would communicate the very same sense of irony to the audience. When she sang a showstopping number called "Diamonds Are a Girl's Best Friend," she did it with a twinkle in her eye and a giggle in her throat.

That said, the lyrics were written from Lorelei's perspective and quite deadpan; the way she coolly prioritized diamonds—her emblem of status and financial security—above kisses, romance, and even long-lasting commitment would have made the copy team at N. W. Ayer & Son proud. Indeed, Dorothy Dignam grabbed the fresh marketing opportunity presented by the show and ran with it. There were still too many dips in the diamond market, too many postwar economic ups and downs, to let a home run like *Gentlemen Prefer Blondes* go unheralded.

She bombarded her contacts with this new rallying cry: diamonds are a girl's best friend. The song, along with Frances Gerety's successful tagline "Diamonds Are Forever," provided two definitive pop-culture statements about the role jewels should play in people's lives. By the time 20th Century-Fox released the movie version of *Gentlemen Prefer Blondes* in 1953, starring Marilyn Monroe as Lorelei Lee, these memorable messages about diamonds had already started to sink in. Gone was Channing's tongue-in-cheek delivery, her defiant glint that signaled she was brighter than the clueless girl she was playing. Monroe was a different kind of Lorelei: stunningly beautiful and unapologetically pert. She, too, was a comedic actress, but when she wiggled on-screen in a pink satin gown and a suite of real diamonds—two necklaces, four bracelets, and earrings so long they dangled down to her collarbone—she made Lorelei look so glamorous that even the most prudent woman in the audience was apt to forget herself. And if that woman—or, perhaps, her increasingly anxious husband feeling his wife's pulse rise—was inclined to shop after watching the film, the song let both know exactly where to go: "Tiffany's! / Cartier! / Black, Starr, Frost, Gorham! / Talk to me, Harry Winston / Tell me all about it!"

No static advertisement, no *Vogue* spread, not even a regal museum exhibit could match the charm that Marilyn Monroe brought to the table when she pursed her red lips and cooed those illustrious names. *That* was star power. That was the sparkle of Hollywood, and the diamond kings wanted to bask in it.

∞∞∞∞∞∞∞∞∞∞∞∞∞∞∞∞∞∞∞∞∞∞∞∞∞∞

The group at N. W. Ayer wasn't the first in the jewelry business to notice Hollywood's unprecedented commercial possibilities. Paul Flato was a well-heeled East Coast jeweler who opened a store in Los Angeles after the director George Cukor invited him to design pieces for his 1938 film *Holiday* starring Katharine Hepburn and Cary Grant. At the end of the film, Flato received a credit for his work—an exceptional accomplishment for a jeweler, and no doubt Flato's own idea, given his talent for self-promotion. He was a caramel-skinned playboy and a bit of a striver; the son of wealthy Texas pioneers, Paul Flato moved to New York for Columbia Business School and found his footing in the moneyed social scene on campus. He joined a fraternity and dated prep-school girls; he frequented parties at the Waldorf and the Plaza. By the age of twenty-seven, Paul Flato had been married twice, was busy running an elegant New York City boutique, and was worth over a million dollars. He plowed through the Jazz Age as the jeweler du jour and coasted past the Depression by serving an international roster of affluent collectors.

When he moved to Los Angeles, he made sure to keep company with the kind of people who attracted tabloid attention. Actresses like Paulette Goddard, Ginger Rogers, Vivien Leigh, and Marlene Dietrich all wore his jewels to high-profile events—like movie premieres, and even the Academy Awards—where they were sure to be photographed. Soon, Flato's creations appeared on Merle Oberon in *That Uncertain Feeling* and on Rita Hayworth in *Blood and Sand*. Even the great songwriter Cole Porter recognized his influence: "If I rattle with clips from Flato / If I battle with frocks from France / It's because I mind my own potatoes / When it comes to love and romance."

Paul Flato was not only an artist but also an early entrepreneur to recognize the power of movie actresses. Actresses were like socialites—except that, from a commercial standpoint, they were even more valuable. First of all, unlike socialites, actresses were necessarily pretty: beauty was the price of doing business (models had to be pretty, too, but they were also typically anonymous). More important: actresses had a platform that even the richest women in America couldn't obtain. What was extraordinary about film stars is that they had the support of the studio system behind them. Studio heads wanted their starlets to get famous, to be publicly recognized and even worshipped, and they were happy to invest their money and clout in order to make sure that this happened.

It's no wonder that, even before the war ended, De Beers had set its sights on Hollywood. In the late 1930s, the sleek and sunny counter to Manhattan was Los Angeles, a playground where all money was new money, all blondes seemed young and beautiful, and palm trees defined the skyline. N. W. Ayer reached out to Margaret "Maggie" Ettinger, a movie publicist with offices on Hollywood and Vine. Ettinger got her start at MGM, and had rebranded an aspiring film-processing client as Technicolor. Her studio background left her with useful connections, and her cousin, gossip columnist Louella O. Parsons, provided her with a privileged entryway into the press. Ayer and De Beers determined it was worth paying Ettinger a salary of $425 a month (the equivalent of about $6,800 today) to single-handedly gild and protect the reputation of diamonds in Hollywood. This meant anything from convincing producers to add diamond sequences to films—a scene in which a man gives his darling a sparkling necklace, for instance—to making sure the stars wore gemstones in their publicity shots. And conveniently, around the same time that Ettinger started working for N. W. Ayer, Parsons also paid particular attention to celebrity engagement rings in her columns: Ingrid Bergman "wore a wedding ring and a diamond engagement ring" while she tearily awaited her husband's return home from the front; "the engagement ring Jeffrey Lynn gave Dana Dale is a star sapphire set with diamonds." Quick, offhand mentions like these weren't particularly titillating, but they supported the industry agenda in two ways: they kept the engagement ring tradition feeling modern, and reinforced the connection between celebrities and diamonds in readers' minds.

Meanwhile, Ettinger advocated for diamonds in the movies at every opportunity. Merle Oberon, who wore all of those Flato pieces in *That Uncertain Feeling*, was also Ettinger's client, and the publicist made sure to promote the jewels as well as the almond-eyed actress when the oddball romantic comedy hit screens in 1941. That same year, a film called *Skylark* premiered starring Academy Award winner Claudette Colbert; it included a lengthy sequence of its protagonist, a frustrated young wife, shopping for diamonds: Ettinger's creative contribution. But perhaps her biggest coup took place when she heard tell of a film in production called *Diamonds Are Dangerous*, in which an upstanding air force captain travels to South Africa and nearly falls under the spell of a beautiful jewel thief but instead follows the call of duty and, in doing so, wins her heart, inspiring her to reform. Somehow, Ettinger managed to convince executives at Paramount to change the film's title to *Adventure in Diamonds*, even though the content wasn't altered. Was a negative title enough to deter potential consumers even though the story itself was romantic and exciting? Clearly, someone who signed Maggie Ettinger's paychecks thought so.

As Ettinger handled Hollywood, Dorothy Dignam remained back in Manhattan, matching the West Coast's efforts. She solicited stories from female celebrities about their engagement rings and then promoted them in articles like "The Day I Got My Diamond" in the April 1959 issue of *Motion Picture* magazine. The vignettes featured high-profile women like Sophia Loren and Cyd Charisse, but Jayne Mansfield's story is the most remarkable for its sweetness—but also for what it implies about her priorities. "The 'man in my life' reached over to hold my hand," she gushed. "He took the third finger of my left hand, kissed it, and slowly slipped on a ring. It seemed to be a plain platinum band. But as I turned around, the most beautiful surprise I have ever had in my life was before me. A stone that sparkled and glittered. It was the biggest diamond I had ever seen." She goes on to say: "Mickey [Hargitay], my husband, gave me that diamond with his love. Our home is not completely furnished. But in the candlelight that Mickey and I often use to light our small dining table, the ring reflects on the walls and all around the room, to bring more light into our happy home."

What's a big diamond among friends? The answer is all but explicit. Who needs furniture, or even electricity, when you have an enviable jewel twinkling on your hand? The day that issue of *Motion Picture* hit newsstands, after getting home from the office, the never-married Dorothy Dignam must have poured herself a well-deserved double martini before tucking back in to work—as was her habit—with her own surprise gift. On her twenty-fifth anniversary at Ayer, her colleagues presented her with the perfect present for a woman who loved her job above all else: not a diamond, or even jewelry at all, but an eight-and-a-half-pound Swiss-made portable Hermès typewriter.

◇◇◇◇◇◇◇◇◇◇◇◇◇◇◇◇◇◇◇◇◇◇◇◇◇◇◇◇◇◇◇◇

The year *Gentlemen Prefer Blondes* premiered in theaters, 1953, was a good time for diamonds. That same summer, Princess Elizabeth was crowned queen in a gemstone-studded celebration that captivated citizens well beyond the borders of the British Empire, and Senator John F. Kennedy—a golden boy known for his good looks and famous family—proposed to Jacqueline Bouvier, a fashionable, dark-haired young woman with an equally impressive pedigree. She was spotted wearing a stunningly unique Van Cleef & Arpels engagement ring, with a 2.84-carat square emerald center stone mounted alongside a 2.88-carat diamond, nestled in a bouquet of marquise-cut stones and baguettes. Obviously, it wasn't the typical solitaire, and its unusual style indicated high craftsmanship. The future First Lady became a lifelong Van Cleef & Arpels devotee, frequently wearing its pieces to events she attended with her husband.

When Marilyn Monroe married Joe DiMaggio in 1954 after a whirlwind courtship, he gave her an eternity ring made of thirty-six baguette diamonds. That wedding caused a stir, but it was nothing compared with the news that broke in the winter of 1956, when the star of *Dial M for Murder* and *Rear Window*, Grace Kelly, made a surprising announcement: she was engaged to Rainier III, prince of Monaco. With the upcoming nuptials, she would make the unlikely jump from Hollywood royalty to true-blue princess. Naturally, fans were surprised and delighted to see one of their favorites blessed with such a rare opportunity, and one that felt particularly elusive to royalty-obsessed Americans. Kelly, who came from a wealthy and prestigious Philadelphia family and who, at least according to her media origin story, was once considered by her parents to be the least promising of her siblings, handled the press with the poise and graciousness of a queen. She confessed when she first met Prince Rainier she didn't feel any sparks, but the next time—at her parents' house in Philadelphia—it was altogether different (love at "second sight," she quipped winningly). The prince proposed on New Year's Eve in 1955, after asking Kelly's father for permission. Gladly, she showed off her gifts from Rainier: a gold-linked bracelet with a matching coin from Monaco dangling from it, and a ruby-and-diamond ring from Cartier, the proud symbol of their engagement.

Or maybe not so proud after all. After spending January on the press circuit showing off one engagement ring, Grace Kelly appeared on the scene in February with another one—she'd made a conspicuous swap. The new ring, also by Cartier, was impossible to miss: a 10.47-carat emerald-cut solitaire mounted in platinum, with smaller baguettes on either side. Casual mentions of the flashy new piece were dropped into articles about the princess-to-be; the *Washington Post* ran an update called "Grace Has Two Rings," which explained that the prince had "replaced" his fiancée's first one, even though it hadn't been lost. Rather, she was now the lucky owner of both.

Nothing seemed too extravagant for a socialite—cum—film star—cum—monarch, so while news of a woman with *two* Cartier engagement rings might have raised eyebrows in different circles, Kelly's fans accepted it as just another day in the life of the enviable blonde. As wedding plans went under way, Prince Rainier commissioned even more fine jewelry for his bride: a wedding suite from Van Cleef & Arpels, with matching necklace, ring, earrings, and bracelet, all made of crisp white diamonds and natural pearls. Within the year, he'd named the house the "Official Supplier of the Principality of Monaco," and Princess Grace joined the ranks of women like Wallis Simpson and Jacqueline Kennedy who'd never want for fine jewelry again.

For a spell, only she and Prince Rainier knew the truth about the new engagement ring, and why a perfectly lovely ruby-and-diamond bauble required a fast replacement. But eventually the story drifted into the public's ear: soon after he proposed, Prince Rainier realized he'd made a terrific blunder. Traveling in the States and meeting his fiancée's rich and famous friends, he couldn't help noticing their predilection for large, flashy diamonds. Supposedly, Kelly's own parents were more than happy to confirm his suspicion that, when it came to jewelry, bigger was always better by American standards. Being a prince with money to spare, he immediately contacted Cartier with a request for a more traditional—and more breathtaking—romantic diamond. Privately, the couple made the switch and hardly felt a need to explain it. Kelly's days in Hollywood were numbered, anyway. Soon after the engagement she announced that MGM's *High Society*—a musical remake of *The Philadelphia Story* starring her as the Katharine Hepburn character, and crooners Bing Crosby and Frank Sinatra as her suitors—would be her last film. The *New York Herald Tribune*, when it caught wind of the glamorous vehicle, called it "a bon voyage basket that would be a credit to Tiffany's window."

Playing a Rhode Island socialite, Grace Kelly had good reason to wear her Cartier diamond ring in the movie; not all Hollywood actresses wore their own jewels on film, but some more successful stars, like Mae West (who was known for playing a character called Diamond Lil) and Joan Crawford, did. Given that it was the era of Maggie Ettinger, Kelly also wore her engagement ring in publicity stills for the movie. The studio likely didn't need much convincing on that front; MGM understood that while the original *Philadelphia Story* was beloved and the two male costars were both renowned, much of *High Society*'s draw had to do with the growing allure surrounding the beautiful actress. The ring signified many things, but among them was the reminder that Hollywood was losing one of its brightest stars. This was the last chance for audiences to admire Monaco's soon-to-be princess in the way they'd fallen in love with her: on the big screen.

✦✦✦✦✦✦✦✦✦✦✦✦✦✦✦✦✦✦

The jewelry industry could rest assured that, whether Hollywood's stars knew it or not, they—along with the newspaper and tabloid journalists who wrote about them—were very dependably promoting the diamond agenda. Still, the bosses at N. W. Ayer & Son were much too savvy to let their employees just kick up their heels and put all of their faith in the whims and hot winds of LA.

Accordingly, Ayer had taken its own stab at the moviemaking business, promoting two homegrown short films called "The Eternal Gem" (1945) and "The Magic Stone" (1946). Both made their way around schools and social clubs, but the latter was picked up by Loew's theaters and played for audiences in cities from New York all the way to San Francisco. In response to the success of "The Magic Stone," N. W. Ayer and De Beers made a third movie in 1953 called, simply, "A Diamond Is Forever." Columbia Pictures distributed the twenty-seven-minute color film across 3,500 theaters that year, and then, in 1954 and 1955, arranged for it to air on television. In that time, "A Diamond Is Forever"—which followed a moony young couple as they shopped for an engagement ring, and featured almost half a million dollars' worth of De Beers diamonds—reached over twenty-three million people in a setting more intimate than any movie theater: their own living rooms. Wearing their curlers and slippers, audiences watched as affianced Mary and Tod learned about gemstones from a kindly jeweler, who described everything from the difficulties of diamond mining to the painstaking process of fashioning. "A Diamond Is Forever" hit on all of Ayer's carefully crafted selling points. Diamonds make people joyful; the film opened with Mary scribbling in her diary, wondering how to describe "the happiest day of [her] life," only to realize that "the ring told everything!" It then flashed back to earlier that day at the shop, where the jeweler narrates a tale of hard work and natural treasure, while rare footage of the South African mines, and the diamond cutter's workshop, played on-screen.

With diamonds, N. W. Ayer had the tricky task of highlighting two seemingly opposite features: their inherent glamour and exclusivity on one hand, and their accessibility and widespread appeal on the other. For the time being, Hollywood could be trusted to continue emphasizing the former. But when it came to promoting the democratic aspect of gemstones, the industry had to rely on more creative tactics: ads and informational pamphlets, of course, but also other methods, like these films, which sought to educate potential customers and make them feel comfortable talking about diamonds. Along with Robert M. Shipley at the GIA, the people at Ayer determined that informing Americans about the product was as important as, if not more important than, tantalizing them.

That's where Gladys B. Hannaford came in.

Hannaford had been working in the industry since the mid-1930s, when she got a job with Harry Winston; she'd been the one at the steps of the American Museum of Natural History with a pen in hand, waiting to sign for the Jonker. Twenty years later, a slim, gray-haired woman in cat-eye frames, Hannaford was N. W. Ayer & Son's resident lecturer, driving her own car around the country to give talks like "The Romance of Diamonds" at churches, high schools, clubs, and universities. Hannaford hadn't always dreamed of being a "career woman," as the *Brooklyn Eagle* dubbed her in its "Beauty After 40" column in 1951, but her life had unfolded in a way that surprised her. She certainly hadn't imagined that she would one day travel thousands of miles below the equator to tour the mines of South Africa. Rather, in her younger days, she'd made wholly conventional choices. She married and gave birth to a daughter—before her husband died suddenly and left her a single parent with a child to support on her own. Forced to consider her job options, she remembered the fun she'd had doing community theater, before her responsibilities as wife and mother got in the way. The skills she'd developed onstage served her well in her work for N. W. Ayer & Son. No matter how many times she'd given a lecture and no matter how exhausted she felt, Hannaford played the part of the Diamond Lady convincingly. Audience members always commented on the charm and passion she brought to her subject.

She was certainly quick on her feet. In the summer of 1952, she appeared on a radio show in the Midwest with a live audience of housewives and a question-and-answer format. The very first question was a doozy: "Is a diamond a good investment?" Hannaford surveyed the crowd while considering how she would turn such a complicated economic calculation into a sound bite. Then it came to her and she smiled. "Is a diamond a good investment?" she asked the interviewer back. "Let me ask you where else a young man can invest, say, two hundred dollars and get a woman to cook for him, do his laundry, sew his buttons, be a constant companion to him, and raise a family for him? Where else, but in a diamond engagement or wedding ring?" Gladys Hannaford loved to hear the housewives laugh.

Most of her appearances were considerably easier. Sure, there was the time that a schoolgirl came bounding toward her, excited to show off her "expensive" ring—which Hannaford could tell cost her fiancé far less than the number he'd quoted. But usually her presentations were straightforward. She talked about famous diamonds like the Jonker, the Hope, and the Koh-i-Noor, and even carried replicas with her so that anyone in the audience could experience the thrill of holding something very close to a priceless gemstone in his or her own hands. She showed slides from her trip to South Africa that illustrated the diamond's magical journey from the depths of the earth—and then explained how the rough stones traveled to places like Israel and Antwerp, where they were cut and polished; then over to the Diamond Dealers Club in New York, where the bearded men carried loupes in their pockets and haggled over prices; and then, if they were high quality, finally to Tiffany & Co.'s in-house workshop and its gorgeous Christmas windows. She taught listeners about the 4Cs. Depending on the demographics of the audience or the purported topic of the day's lecture, she was prepared to chatter about everything from cutting-edge jewelry trends to the history of diamonds in art and literature. Always, the point was to stimulate interest in gemstones, to make sure they seemed deeply important yet still unintimidating.

N. W. Ayer & Son particularly liked it when she presented her program in schools. There, among the impressionable young students, Hannaford could single-handedly influence the very people who were poised at the threshold of graduations and engagements, and thus would soon become the next generation of diamond buyers and recipients. Indeed, Hannaford was always careful to toe the party line. She knew where her bread was buttered—and what other women her age were lucky enough to live such an adventurous life? When asked, at a session of the Fifty-Second Annual New York State Retail Jewelers' Convention in Binghamton, whether wide wedding bands—a new style, which was being promoted by some jewelers, and which Hannaford knew that the people at her agency loathed because such bands didn't leave enough room for a diamond engagement ring—were the future of bridal jewelry, she was sure to answer, "Absolutely not!" No matter how many significant stones and details she discussed, she was sure to leave audiences with a clear, concise takeaway, as she did when she spoke to a group at Benson High School in Omaha, Nebraska: "The most important diamonds are those worn on the third finger, left hand of American girls."

In 1960, N. W. Ayer & Son commissioned a survey of twenty thousand American households to determine the present effectiveness of the campaign that had been running for two decades. The findings were remarkable: not only did 80 percent of brides receive diamond engagement rings in 1959—an all-time peak—but the carat weight of those stones seemed to be steadily increasing. Here was proof that the ritual of proposing with a diamond ring had been successfully instated in contemporary culture. More than that—it had, according to National Family Opinion Inc. (the company that conducted the study), hit a kind of critical mass. "When we look at the characteristics of the women who have not received a diamond engagement ring at the time of marriage," the researchers determined, "it seems clear that forces beyond the influence of promotion must have been at work to produce this situation—lack of income and education, age, previous marital status, quick marriage with no engagement ring, etc."

In other words, the only way to increase product sales at this point would be to encourage more weddings. Luckily, the early to mid-1960s represented an exciting social watershed for jewelers. As the *New York Times* reported in January 1962: "Those born during the post-war baby boom will be reaching the age for high school graduation in 1963 and, subsequently, for engagement and marriage. Coupled with the trend toward earlier marriages—67 per cent of the women who marry do so by the time they are 24—this means sales in every facet of the trade."

◇◇◇◇◇◇◇◇◇◇◇◇◇◇◇◇◇◇◇◇◇◇◇◇◇◇◇◇◇◇◇◇◇◇◇

Thanks in part to the power of Hollywood, 1961 was a terrific year for Tiffany & Co. It wasn't just that company profits rose by a whopping 40 percent since the year before—which they did—or that it had a new leader who was determined to keep up the trend. In the fall, Paramount released a film that put the store's name in every American's mind, and on the lips of one particularly stylish actress: the beloved Audrey Hepburn.

That film, of course, was *Breakfast at Tiffany's*, based on Truman Capote's 1958 novella. Hepburn's Holly Golightly was a stunning young woman who was gripped by two utterly contradictory impulses: she wished to be independent, but at the price of being "kept" by a very rich man. For Holly, the place that felt most like home was the Fifth Avenue store that gave the film its name. "I'm crazy about Tiffany's," Holly Golightly tells her mystified neighbor Paul Varjak upon their first meeting, "the quietness and the proud look of it. Nothing very bad could happen to you there."

The opening scene of *Breakfast at Tiffany's*, which served as the opening of the theatrical trailer as well, shows a close-up of the store's name hung in stately metallic letters over the door. Holly arrives in a yellow cab on an otherwise empty street, wearing her clothes from the night before—the little black Givenchy dress and oversize Oliver Goldsmith sunglasses that she made famous— and stops in front of Tiffany's, where she enjoys a breakfast pastry and coffee, captivated by the jewelry windows. The New York flagship appears once more in the movie, when Holly and Paul are on their first real date. They stroll past Tiffany's glass cases, looking for something to buy for ten dollars. The room, on the ground floor of the mansion with its high ceilings and wood-paneled walls, looks grand. There, Holly confesses that she doesn't "give a hoot about jewelry—except diamonds, of course," gesturing toward a particularly large and elaborate specimen. Her eyes, however, are bigger than her wallet. The store clerk obliges the pair, if a bit sardonically, when they ask about Tiffany's least expensive gifts. He suggests a silver telephone dialer, but they ultimately decide to have a Cracker Jack ring engraved.

The real Tiffany & Co. welcomed its role on the big screen and cooperated with Paramount, opening its doors to a film crew for the first time in history. In appreciation, the studio hosted a private screening for employees before *Breakfast at Tiffany's* came out in theaters. Tiffany & Co. also endorsed the film by lending its renowned 128.54-carat cushion-cut fancy yellow diamond to Audrey Hepburn to wear for appearances and photo shoots leading up to the premiere. Since 1878, when the firm acquired the enormous stone, it had functioned as something of a mascot for the brand—always sparkling against a backdrop of regal blue velvet, but never against skin. However, that changed when Walter Hoving of the Hoving Corporation took over as chairman in 1955, in a sale that marked the end of the Tiffany family's long era of leadership. Hoving wanted to put the Tiffany Diamond on the market. To that end, he asked recently appointed vice president and jewelry designer Jean Schlumberger (*SHLOOM-ber-zhay*) to devise a setting for the canary gem that could intelligently be considered a wearable piece—it was, after all, almost three times the size of the Hope.

Schlumberger, a French jeweler known for his sculptural, often whimsical jewels, was up to the challenge. He sketched and experimented and, in the end, conceived of a "wardrobe" of three settings, made almost entirely of white diamonds and platinum to complement the golden stone. The first and most famous was a swirling ribbon clip that surrounded the Tiffany Diamond like the mane of a lion. It could be worn on its own as a brooch or in the ribbon rosette necklace: a collar that rested just above the décolletage and continued the pattern of curled, crisscrossing strands. Second was another brooch clip with feathery flourishes like angel wings shooting out from it. Last was a naturalistic bracelet with the Tiffany Diamond entwined in a nest of flowers and vines. In November 1956—just one month after one of Schlumberger's pins appeared on the cover model for the October issue of *Vogue*—the new settings were featured in a *Vogue* spread, along with news of the suite's $583,000 price tag (about $5 million today). Tiffany & Co. also put a photograph of the naked diamond, without any setting at all, on the cover of the 1959 Blue Book.

The Schlumberger necklace looked perfect on slim, elegant Hepburn, who tried it on for photographers after shooting the opening sequence of *Breakfast at Tiffany's*. She never wore it in the film, but it did have a cameo: it's the piece Holly points out to Paul Varjak on their date, propped up in its own display. Yet for all the attention Hepburn brought to the Tiffany Diamond, she wasn't actually the first one to wear it. The honor belonged to Mrs. Sheldon Whitehouse, the chairman of the Tiffany Ball in Newport, Rhode Island, in the summer of 1957. In the weeks leading up to the benefit, Tiffany & Co. announced that one woman would be chosen to model it, but her identity would be left a mystery. Immediately, society women started writing in to volunteer themselves. The Saturday before the ball, two armed guards traveled to Rhode Island to deliver the stone, hanging from a chain of white diamonds. Mrs. Whitehouse paired it with a gown, white gloves, and her own earrings and bracelet. As she entered the Marble House, the East Coast's most illustrious women craned their necks to see who was lucky enough to be wearing the famed Tiffany Diamond, and two guards trailed Mrs. Whitehouse as she strutted from guest to guest. The jewels on display were enviable, but none quite as eye-catching as the yellow stone suspended from her neck.

Walter Hoving flattered his hostess that night. However, Mrs. Sheldon Whitehouse deflected the compliments, telling the *New York Times*, "It might have been more attractive to have a young woman wear it." A woman like Audrey Hepburn, perhaps.

Hoving prided himself on his strong opinions. When he took over at Tiffany & Co., he made it clear that, above all else, the brand should be a paragon of good taste. "We won't sell diamond rings for men because we don't like it," he told a reporter for the *New York Times* in 1965. "We don't like silver jewelry either. People have to get used to the Tiffany point of view in most cases and we will not change for a particular community." If that staunch perspective—

the customer is sometimes wrong?—took some people by surprise, Hoving was quick to remind them that, before he came into the picture, Tiffany & Co. was stuck in a slump. In 1955, business was down and the merchandise, slow to move off the shelves, was reminiscent of a dim, dusty era. The firm had been wedged in an uncomfortable position, trying to honor its prim legacy on one hand and scrambling to anticipate the desires of a new generation on the other. One of Hoving's first moves as chairman was to host a fire sale, shocking customers who never thought they'd put the words "clearance" and "Tiffany's" in the same sentence. No matter—he wanted to rid the shelves of antique silver, old-fashioned stationery, and fusty ceramics. Anything that didn't sell, he had tossed into the trash.

Hoving hired new designers like Jean Schlumberger, luring him to the United States with the promise of a branded Schlumberger boutique within the Tiffany store. But perhaps his most important—and controversial—change had to do with lowering the Tiffany & Co. price point. Yes, the store would always stock items that appealed to the wealthiest and most discriminating customers. However, Hoving also saw value in carrying lower-ticket items that would draw in a younger, less affluent crowd. As a company spokesperson told *Newsday* in 1960: "Our younger customers are becoming educated to good taste. . . . Now, they're buying Tiffany highball glasses at $12 a dozen. Someday, we expect they'll buy others that we sell—at $540 a dozen." In other words, highball glasses, like a diamond engagement ring, could act as a gateway, too. Around the same time, Tiffany & Co. updated the style of its print advertisements to be simultaneously glossier and more playful. These ads did away with the ornate copy, elegant fonts, and rows upon rows of diamonds that characterized earlier campaigns. The new ads were entirely different: sexy and bold. In *The New Yorker* in 1963: a giant emerald-cut solitaire caught in a spiderweb against a plain black background, no text. In *Houston Town & Country* two years later: a diamond in a pill bottle, with the copy "Take one before proposing." Another 1965 ad: a diamond ring looped around a cat's tail, with the line underneath it: "*You* only live once."

If anyone disagreed with Walter Hoving's methods, it remained difficult to argue with his results. Tiffany & Co.'s profits rose year after year, making longtime Tiffany devotees—or at least shareholders—a little less nostalgic for its former reputation as another fussy high jeweler. Not only did the price of shares skyrocket, but Tiffany & Co. also expanded significantly, popping up in markets like San Francisco, Houston, Beverly Hills, and Chicago. Tiffany left the task of channeling the Old World to the European houses of Cartier and Van Cleef & Arpels, and in exchange reinvented the brand and opened its doors to an eager demographic: as *Newsday* cheekily put it, "a compact car clientele." Meanwhile, Harry Winston chugged merrily along with his collection of big stones, his booming wholesale diamond business, and the occasional publicity stunt. In 1958, he generously donated the Hope Diamond to the Smithsonian in Washington, DC, in the interest of building an American public gemstone collection that would rival Great Britain's. Once again, as he did with the Jonker, he had the diamond sent by US mail, paying a total of $145.29 ($1,190 today) for postage, insurance, and a carrying surcharge,* and drumming up a ton of press—all without showing his face, of course.

By the mid-1960s, the big four jewelers—Cartier, Tiffany & Co.,

. . . .
*After the fact, the USPS noted a tabulation error and billed Harry Winston an additional $12.35, bringing the grand total for shipping the Hope Diamond up to $157.64, or $1,297 today.

Van Cleef & Arpels, and Harry Winston—had developed their own personalities and, in doing so, had each cultivated a base of loyal customers. But soon, wealthy Americans would have yet another house angling to grant all their fine jewelry—related wishes. Would the United States welcome a fifth Fifth Avenue jeweler? Perhaps—this house, called Bulgari, had the advantage of untapped Mediterranean exoticism on its side.

The Winners

United States and Italy

Jewelers and advertising executives weren't the only ones who noticed that Hollywood actresses were quickly becoming America's answer to European royalty. It certainly wasn't lost on Sara Sothern, a onetime aspiring Broadway starlet and devoted mother of two living in a picturesque suburb of London. Although her daughter, Elizabeth, was still quite young, it was apparent to Sothern that she would grow up to be somebody special. The child was mild-mannered with a sweet, almost aristocratic bearing, and she was beautiful, with inky black hair and eyes so intensely blue they appeared purple in some lights. Sothern, an American transplant married to an art gallerist, liked to imagine that one day Elizabeth would marry a titled English gentleman. That way, she'd never want for anything and, almost more important, she would never have to fret about class and fitting in with society people—the way her mother always did.

Unfortunately, those dreams were dashed when the war broke out in Europe and the family decided to move back home to the States. Not only was England of the late 1930s dangerous, but it was also inhospitable to a businessman who made his entire income selling art. And so Sara Sothern's husband—a man named Francis Taylor—moved the family to the sunny California district of Beverly Hills. There, Sothern revised her ambitions for her daughter: even if Elizabeth wouldn't grow up to be a duchess or viscountess, she could still become a star. While it wasn't the original goal, the advantages of this new plan were obvious. In England, Elizabeth would have to wait until she reached marrying age in order to live out her mother's fantasies. Now, in Los Angeles, there was opportunity for immediate gratification. Hollywood had a need for charismatic child actors. In fact, on the boat trip from London, mother and daughter had watched *The Little Princess*, starring Shirley Temple. To Sothern, Temple had nothing on her daughter, and at just nine years old, Elizabeth was even younger and more stunning than the reigning child actress.

The Taylors mined their social connections and brought Elizabeth to a meeting at Universal, where studio executives confirmed Sara Sothern's suspicions: yes, there was something different—and inherently marketable—about this girl. Elizabeth was offered a contract, but just a year later, she had failed to live up to her potential on-screen and was dropped. The setback brought the Taylors to the doorstep of Metro-Goldwyn-Mayer, where Louis B. Mayer gave Elizabeth a seven-year deal and, a few years later, the script that would change her life. Elizabeth had always been an animal lover, and she'd ridden horses on the family's sprawling property outside of London. When the lead role in *National Velvet* became available, even the girl herself—now thirteen—knew it was a perfect fit. She bonded with the gelding cast as the Pie, while her mother arrived on set every day, ready to stand on the sidelines and give her advice about how exactly she should play the role of headstrong Velvet Brown.

National Velvet helped Sara Sothern achieve her dreams, which over time had become her daughter's, too: Elizabeth Taylor was a bona fide star. And yet fame didn't make the young girl happy. As an adolescent, Taylor began to resent the sacrifices she'd made to live the life of a studio actress. She hadn't gone to school in years, had barely any friends her own age, and—if Mr. Mayer had any say— she would never attend a prom. For the first time, she resented her mother's tireless involvement in her career. She sparred with studio executives and challenged Mayer to fire her. He didn't— which only reinforced Taylor's mounting suspicion that the people at MGM needed her more than she needed them.

Taylor had natural talent as an actress, but she never trained and she didn't take the work all that seriously, and as she pushed further into her teenage years, she had to seek out alternative means to boost her self-esteem. She found what she was looking for in men. She was eager to find a husband—her mother had always made marriage sound like the true estimation of a woman's worth— and to start her own family, which in Taylor's mind would create a necessary boundary between her childhood, controlled by Sothern, and an independent adult life. Having experienced few things outside the world of movie magic, she had a naive, almost fairy-tale impression of romance. When, at the age of eighteen, she met

Conrad "Nicky" Hilton Jr., she knew instantly that he was perfect; handsome, charming, and rich, Hilton was like a storybook prince. They got married after four short months of dating.

The relationship proved to be eye-opening for Elizabeth Taylor. On their long European honeymoon, Hilton revealed his true colors: he was a drunk and a gambler who took his anger over his losses out on his new wife. After a high-profile wedding—Taylor was a film star, after all—she was humiliated, not to mention devastated, to have to file for divorce six months later. Her failed marriage to Hilton was followed by two more terrible disappointments. She married Michael Wilding, an actor almost twice her age, and had two sons with him before her mounting fame and success—in contrast to his own stalling career—came between them. She then met Mike Todd, a Hollywood producer whose big personality and bawdy sense of humor made him a good match for the gorgeous spitfire. Aware that Taylor's two previous husbands had been rich, and had given her the kinds of gifts expected of men of their rank, Todd made sure to outdo them both: he proposed with a 29.4-carat emerald-cut diamond ring. When the actress got pregnant, he surprised her with a Cartier suite made up of a stunning ruby-and-diamond necklace, matching earrings, and a bracelet. Friends delighted that Taylor had finally found her soul mate, and she gave birth to their daughter, Liza. However, Todd was killed in an accident on his private jet. Taylor was heartbroken, and additionally distraught over the fact that she was meant to be on that plane with Todd but a bout of pneumonia had made it impossible for her to fly.

By her late twenties, Taylor still felt that she was waiting for her prince, even though her trials had left her cynical about the possibility of real long-lasting love. Her experience of suffering didn't deter her from dating, but it made her more reckless and, arguably, more selfish. At every chance, she seized the potential for romance wherever it appeared to her. Anyway, she had something else in her life to fill the gaping hole left by three failed relationships. Elizabeth Taylor had jewelry, which she was happy to admire and which required almost nothing—except an outlay of money—in return.

Maybe marriage didn't always last forever, the way it had for her parents. She knew of something else that could promise her "forever," and she could hold it in the palm of her hand: a rare, shimmering diamond.

◇◇◇◇◇◇◇◇◇◇◇◇◇◇◇◇◇◇◇◇◇◇◇◇◇◇◇◇◇◇◇◇◇◇◇◇◇◇

By the time Taylor ended up in Rome on the set of the epic film *Cleopatra*, she'd broken up the marriage between two of her closest friends—Eddie Fisher and Debbie Reynolds had served as best man and matron of honor at her wedding to Mike Todd—and married Fisher, despite the considerable hit her reputation had taken in the tabloids. Taylor had become the nation's most notorious home wrecker, and she was castigated by the press until yet another tragedy struck. It happened during the filming of *Cleopatra*: a movie Taylor hadn't even wanted to do when she first got the call from Fox. She read the script and it was awful. But producers were hounding her, and in order to shut them up, she made a ridiculous offer: she'd play the title role for a salary of no less than $1 million plus 10 percent of the total gross. To her shock, they accepted her terms.

That's how Taylor and Fisher ended up on a troubled movie shoot in the United Kingdom, where cost overages and weather delays threatened the entire production. Then, the leading lady came down with Malta fever, a bacteria illness transmitted most often through unpasteurized milk, which required her to be hospitalized. Suddenly, the very same media outlets that had demonized her were encouraging their readers to pray for her swift recovery. The director quit, and although Taylor's health eventually improved, she was too weak to resume work right away.

Eventually, filming on *Cleopatra* was rescheduled with a new director and a new leading man, but this time the location was Rome. In the years since the war, Italy had successfully transformed its reputation. During the darkest hours of Mussolini's reign, the culture was so reviled that Italians living in the United States and Canada were forced into internment camps just for "speaking the enemy's language," and comically square-jawed, thick-lipped villains became omnipresent in English-language books and films. Yet after Italy's defeat, citizens of Allied countries allowed themselves to explore the mystique of the former rival. The movie industry helped. Studios started filming in Rome to take advantage of lower production costs. Bustling with American movie crews and California's most glamorous stars, Rome of the midcentury was dubbed "Hollywood on the Tiber."

There, famous actresses like Elizabeth Taylor fell in love with the city's food, its fashion, and—most dangerously for her costar and new lover, Richard Burton—its jewelry. By the early 1960s, diamond connoisseurs living in the States heard word of an up-and-coming jeweler: Bulgari.

Of course, Bulgari was well established in its home country, but it took women like Taylor and America's sophisticated ambassador to Italy, Clare Boothe Luce, who also fell head over heels for the house, to carry its name across the Atlantic. In 1949, in its "Italian Handbook," *Vogue* referred to Bulgari as the "Cartier of Italy." Tourism to Rome was on the upswing, thanks to films like *Roman Holiday*, starring Audrey Hepburn, and Fellini's *La Dolce Vita*, both of which used the city itself, with its moody narrow streets and ancient architecture, as a character. Bulgari, on the charming Via dei Condotti, was a destination spot, where casual shoppers might rub elbows with the crème de la crème: people like the Astors, Woolworth heiress Barbara Hutton, Babe Paley, and even the Kennedys, who all had the money to spend on extremely high-end jewels.

The store had been there since 1905, when the native Greek artisan Sotirios Voulgaris, born of a family of silversmiths, moved his stock of bric-a-brac, antiques, and jewelry from a smaller location a few doors down. Voulgaris had expatriated to Italy with his father at the age of twenty after a series of violent incidents between the Turks and the Christians made Greece too unstable for the family to remain there. Sotirios opened his first shop on Via dei Condotti in 1894 when he—by then a nattily dressed father of six with a round face and a broad mustache—straightened out the Hellenic curves of his name to become Sotirio Bulgari, proprietor. His store, then called S. Bulgari, was popular among foreign tourists, particularly ones from America and the United Kingdom, and Sotirio had an instinct for how to draw in more customers after a childhood spent watching his father peddle silver. When he opened at the new, bigger location, he labeled S. Bulgari an "Old Curiosity Shop," referencing the Charles Dickens novel. If English-speakers wanted to give him their lire, then Sotirio would make sure to let them know they were more than welcome.

He died in 1932 between the two world wars, when retail in Europe was precarious. Still, his two eldest sons, Costantino and Giorgio, who had trained diligently under their father, chose this as the moment to expand. Before Sotirio passed away, they'd transformed Bulgari from a quaint corner store to a more serious jeweler, and they wanted to build a shop that better represented their

hard-earned status. The renovated location had crystal chandeliers and marble columns. Certainly, the brothers had both committed to the family business, in ways that suited their individual
interests and talents. Scholarly Costantino, who was expert in the
decorative arts, stayed faithful to the Bulgari legacy and traveled
the world, learning more about silver. On the other hand, Giorgio
was more business-minded. He traveled, too, but his focus was on
gemstones. He was especially influenced by Parisian jewelers like
Boucheron and Cartier and, in studying them, earned himself the
title in Italy of "Jeweler to Kings."

Kings make great customers, but if a house wanted press in the
late 1950s and early 1960s it was better off catering to film stars.
By this time, Bulgari was under the control of Giorgio's three sons,
Giovanni (known as Gianni), Paolo, and Nicola. Handsome Gianni,
the eldest, inherited his father's love of gemstones and his talent
for design. He was also a natural salesman.

◇◇◇◇◇◇◇◇◇◇◇◇◇◇◇◇◇◇◇◇◇◇◇◇◇◇◇◇◇◇◇◇◇

At the age of thirty Elizabeth Taylor already had a mature collection
of jewelry—flaming rubies from Mike Todd and daintily strung diamonds dating all the way back to the belle epoque—and any man
who tried to impress her had to make a grand gesture in order to
step out of the shadow of three rich former husbands. Asking men
for baubles was half challenge, half flirtation; when Todd was pursuing her, she told him she'd go out with him—but only if he bought
her a diamond-and-emerald pin. She wasn't just a passive recip-
ient, either; she had been honing her eye for jewels since the age
of twelve, when she saved up her MGM contract money to buy her
mother a real diamond, gold, and sapphire brooch. Her mother was

both surprised and touched; since then, Taylor had never forgotten the way that beautiful gemstones could make a person feel lovely and appreciated.

One afternoon, the glamorous actress with glossy black hair and violet eyes walked into the Bulgari store. *Cleopatra* was filming, and although the outrageous production still suffered from budgetary setbacks, the biggest problem had to do with the rumored affair between Taylor and Richard Burton, who was playing Mark Antony, a blue-eyed, squared-jawed actor considered a successor to the great Laurence Olivier. The pair spent hours in each other's trailers, and supposedly kept their love scenes going long after the director yelled cut. Their romance might have provided a welcome jolt of publicity if both stars hadn't been married, and in fact the cuckolded Eddie Fisher was still living in Rome with his wife. The paparazzi—who, as an industry, had only recently emerged as a by-product of Italy's film renaissance—loved a good cheating scandal, and the local press swarmed, determined to prove that Taylor and Burton were betraying their spouses together. When Fisher packed up and left Italy, tabloid writers considered their suspicions confirmed. After that, the besotted stars were even more brazen. They delayed *Cleopatra*'s schedule by disappearing for hours and hours together. They were photographed flirting in their off-hours, performing for the telephoto lenses. One afternoon, generous Burton offered to buy his lover a present. Taylor knew exactly where she wanted to go. She led him to the Via dei Condotti.

In *My Love Affair with Jewelry*, Taylor reminisces that the best thing about working in Rome was "Bulgari's nice little shop." Later, Richard Burton quipped that "Bulgari" was the only Italian word she picked up while on the set of *Cleopatra*. Living in Rome, she developed a friendship with Gianni Bulgari, who happened to be on hand the day Taylor and Burton arrived. Burton knew his audience: he gave his girlfriend a budget of $100,000 (about $777,000 now). But Gianni—a jet-set playboy with highborn grace and manners— knew his customer, too. Slyly, he slipped into the safe and returned with a pair of earrings so small they were sure to disappoint the woman whose late husband had proposed with a diamond the size of a golf ball. When Taylor appeared underwhelmed, Gianni stayed firm: "That's $100,000."

Sure enough, Burton urged him to bring out something else.

After she rejected the decoy earrings, Gianni reached deeper into his treasure chest and started showing off the best that Bulgari had to offer. Only recently, the firm's design direction had shifted, and Gianni knew that Elizabeth Taylor was precisely the kind of woman who would be bold enough to wear the season's new styles. In the 1950s, when fashion was feminine and delicate, Bulgari was known for its tremblant brooches: pristine little floral sprays inspired by pieces from the eighteenth century. The word "tremblant" came from the French *en tremblant*, which means trembling, and Bulgari set white diamonds and colored gemstones on tiny springs so that they shivered in place like leaves in the wind. These jewels were quite popular with prominent women, and Taylor already had her own emerald-and-diamond version, given to her by Eddie Fisher before their marriage fell apart. The tremblant remained a signature for the house through the late 1950s and early 1960s, but in that time Bulgari also started working with heavier, flashier pieces, like colorful necklaces that combined classic white diamonds with other rich, bright stones, often cut using an old technique called cabochon. A cabochon ruby, for instance, is polished rather than faceted, so its surface is smooth and round like a red bead. When scintillating diamonds were mixed with more opaque, glassy stones, the resulting creation was both naturalistic—as if the colored gems had been discovered by chance in the sand at the mouth of the ocean—and elegant: without a doubt, the work of the finest jeweler.

At the store, Gianni revealed "green flames" of emeralds and a ruby-and-diamond necklace that made Taylor's heart flutter. When Gianni confessed, only faux apologetically, that it cost over a million dollars, Burton refused and Taylor backed off—she had the Cartier ruby-and-diamond necklace from Mike Todd, anyway. Finally, Taylor decided she wanted emeralds, and she tried on two emerald-and-diamond necklaces—one larger and one smaller— each worth well over $100,000. Ultimately, she decided she liked the smaller one better, in part because it had a detachable emerald-and-diamond pendant that could be worn alone as a brooch. "It's really like getting two pieces in one," she assured her benefactor cheerily, as he doubtless resigned himself to a much more substantial purchase than he'd originally planned. He couldn't help it—he was totally smitten. Later, they'd agree that the detachable pendant, at 23.44 carats, would stand in for an engagement ring. The matching necklace—which featured sixteen emeralds of various sizes surrounded by many more white brilliants, pear shapes, and marquise diamonds—would eventually be described as a wedding present.

After all that she'd been through, Taylor must have felt just a little bit untouchable. Burton agreed to leave his wife of fourteen years for her, and even though the Vatican denounced her as an adulteress, the fans remained loyal. They made their voices heard during the filming of a large crowd scene in *Cleopatra*, in which she stood on a balcony as her subjects shouted praise for their queen. Although producers feared the actress might be assassinated on account of all the negative press, the extras actually expressed their support: they yelled out not for Cleopatra but for "Liz."

Even after the movie premiered to bad reviews, Taylor had two souvenirs from her time in Rome: her dashing new husband and her Bulgari jewels. She wore them throughout her life, along with a matching pair of emerald-and-diamond earrings—a thirty-second-birthday present from Burton—as well as a Bulgari emerald-and-diamond bracelet and cocktail ring. In the 1963 British film *The V.I.P.s*, which was released after *Cleopatra* and starred Taylor and Burton as estranged spouses, Taylor played a glamorous movie actress and wore her own Bulgari tremblant brooch, her emerald earrings, and her engagement pendant matched with an off-shoulder white gown. She wore the engagement brooch as often as she could: pinned to her dress for her wedding to Richard Burton in 1964, of course, but also at the various stylish events that filled the couple's lives, like the first night of Sammy Davis Jr.'s engagement at the Copacabana in New York or the premiere of *Lawrence of Arabia* in Paris. In 1967 she wore the full necklace to the BAFTA Awards, where she and Burton were both honored for their performances in *Who's Afraid of Virginia Woolf?* By then, a handful of jewelers had already discovered that awards shows made for terrific press opportunities. Elizabeth Taylor, as a wealthy woman in her own right and the lucky beneficiary of countless love tokens, had plenty of baubles to choose from when she got dressed up for these kinds of affairs. But for other actresses, like the seductive Italian Gina Lollobrigida, an ongoing arrangement with the right house could be a very good— and mutually beneficial—thing.

The Italian actress Gina Lollobrigida actually resembled Taylor, with dark hair and fine features, but her exotic, almond eyes gave her the look of a wildcat, which enhanced her image as a sex kitten. She had a body like a pinup drawing and flaunted it, as if doing her own small part to help break the world's negative wartime associations with her native country. On-screen, she wasn't as charismatic as Italy's other screen goddess, Sophia Loren, but she brought the same sensuality to her films. Lollobrigida was a great Bulgari customer. But her relationship with the house also had terrific reciprocity; for every piece she bought, she borrowed another, which she then wore to her movie premieres and other high-profile appearances.

The first Academy Awards ceremony was held in 1929, a quiet affair at the Hollywood Roosevelt Hotel where the honorees knew in advance that they'd won. It wasn't until 1941 that the Academy started guarding the results, and this element of surprise brought additional public interest to the event. Paul Flato, with his instinct for self-promotion, began lending out his jewelry around that time to actresses in the running for the most prestigious titles. When Ginger Rogers won Best Actress for *Kitty Foyle* in 1940, she accepted her statue wearing a unique floral diamond cascade necklace that rested around her neck and stayed open in front, culminating in a splash of white stones. She wore a complementary Flato bracelet and ring as well. However, Harry Winston, not Flato, tends to get the credit for "inventing" the modern-day practice of lending jewels to high-profile stars for awards shows. In early 1944 Winston received an invitation from MGM producer David O. Selznick to accessorize his leading lady Jennifer Jones for her very first Academy

Awards ceremony. Selznick was betting that she'd win Best Actress for 1943's *The Song of Bernadette*, and wanted the twenty-four-year-old Jones to look the part of a victor. The night of the Oscars, she wore a black ruffled dress with a cherry-red manicure, and sparkling Harry Winston jewels. She won, and as she gripped her Oscar, the cameras closed in on the innocent, joyful face that Fox's film about the misunderstood teenage saint had made famous.

In 1953, when the Academy Awards were first televised, the benefits of a relationship between jeweler and honoree were all the more obvious. By that time, there was a second big awards show where jewelers could essentially use actresses to model their creations in a most glamorous setting: the Golden Globes. Founded by a group of foreign correspondents based in Los Angeles, the Globes began in 1944 as a low-key ceremony on the 20th Century-Fox lot, where Jennifer Jones was once again selected as the best female performer of the year. In 1945, the show—which was conceived as a public forum for this group of overseas journalists, eventually called the Hollywood Foreign Press Association, to mete out their views on the year's strongest films—found a more fashionable setting at the Beverly Hills Hotel.

At the 1961 Golden Globes, Gina Lollobrigida was nominated for a Henrietta Award, which had been added in 1950 to recognize the best male and female performances from international film. Lollobrigida looked radiant as she stepped out in Los Angeles, the glitzy town she'd once rejected despite some compelling invitations to move there, because she loved Rome too much. That night, she had her wavy hair pinned up, and wore a textured white gown with a revealing scoop neckline. The jewels were all Bulgari: a pair of cascading diamond earrings, a geometric bracelet, and, most famously, an exquisite diamond necklace made up of round bril-

liants haloed by half-moon baguettes, in the art deco style. Lollobrigida wore the same necklace in her hair as a tiara at Italy's David
di Donatello Awards in 1969 when she won Best Actress (tied with
Monica Vitti) for her performance in *Buona Sera, Mrs. Campbell*,
and again in close-up publicity stills that showed off her striking
Roman profile—as well as the piece's remarkable craftsmanship.
The bracelet and necklace combination became so firmly associated with Lollobrigida that when she put her jewelry up for auction
with Sotheby's Geneva in 2013, it fetched $783,851, hundreds of
thousands of dollars above the initial $500,000 estimate.

Lollobrigida was often linked to Gianni Bulgari in the tabloids, but her crush was largely unrequited; as he told the *Boston
Globe* soon after *Vogue* proclaimed him an eligible bachelor, he
preferred blondes. Still, he didn't stay in one place long enough to
invest in lasting love. After falling hard for racing cars, he learned
to fly planes and traveled across the world sourcing gemstones and
other design inspirations. He had his sights set on international
expansion, and in 1967, Bulgari ran its first print advertisement
in American *Vogue*. Four years later, he got his wish for a New York
store when Bulgari opened a branch at the Pierre Hotel, an impos-
ing building at the foot of Central Park. There, the house would
sell its flower brooches, Roman coin necklaces, and curling snake
bracelets—the snake becoming as iconic for Bulgari as the panther
was for Cartier. The boutique opened with a luncheon for two hundred, including the competition. Claude Arpels and Walter Hoving
attended, the latter expressing his goodwill to the *New York Times*:
"I'm just tickled to death [that Bulgari is in New York]. We'll all do
much more business because they're here. This makes New York
the center of the jewelry business in the world. We have everything
now that Paris and Rome do."

That day in December 1971, the best jewelers in the world celebrated their success. Diamonds and other gemstones were prized
everywhere from Hollywood to Rome to Bombay and back again.
George Balanchine—one of the greatest choreographers of his time
and a close friend of Claude Arpels's—honored their magic in 1967,
creating a ballet called *Jewels*, with three separate movements:
"Emeralds," "Rubies," and "Diamonds." A few years after Bulgari
opened in New York, Andy Warhol announced that he "always visit[s] Bulgari, because it is the most important museum of contemporary art." Jewels were no longer just romantic tokens or status
symbols—now they were significant pieces of contemporary high
culture. They had, in the eyes of the sellers, earned their price. "We
like to think of our jewelry as art," Gianni Bulgari told the *New York
Times* at the opening in New York, "and since Americans pay hundreds of thousands of dollars for art, there is no reason why they
shouldn't pay that for jewelry." It was an edict that Elizabeth Taylor
and Richard Burton, when they returned to the States, took to heart.

Portrait of Cornelia Bradley-Martin at the Bradley-Martin Ball.

Jacob & Co. unveiled the 260-carat Billionaire watch at Baselworld 2015.

Charles Lewis Tiffany, once known as the "King of Diamonds."

Maiden Lane, New York's original diamond district, conveniently abutted the seaport.

Cecil Rhodes, founder of
De Beers, was nicknamed
the Colossus.

The Premier mine in South Africa produced the Cullinan Diamond
and the Taylor-Burton Diamond.

Eleanor Roosevelt's engagement ring is an early example of the Tiffany Setting.

Agnès Sorel is believed to be the model for the Madonna in Jean Fouquet's *Virgin and Child Surrounded by Angels.*

A young Evalyn
Walsh McLean
poses in her two
most famous
jewels: the
Star of the East
(in her hair)
and the Hope
Diamond.

The spellbinding Hope Diamond, believed by many to be cursed.

(ABOVE) One of three brothers, Pierre Cartier moved to New York from France to open Cartier's first U.S. store.

Shirley Temple posed with the rough Jonker Diamond as a part of Harry Winston's publicity campaign for the stone.

Marilyn Monroe (as Lorelei Lee) convinced audiences that
diamonds were, in fact, a girl's best friend.

The House of
Jewels was a
popular exhibit
at the 1939–1940
World's Fair in
Queens, New
York.

She's Engaged to a member of the Royal Canadian Air Force

She's Lovely!

HER RING—an upraised center diamond flanked by smaller diamonds on intricate design in gold.

FRANCES KING, of Poughkeepsie, N. Y., of the old Hudson River family—another lovely Pond's bride-to-be. Her engagement to H. Paul Richards, of the R.C.A.F., was announced by her mother, last May.

Pretty as a picture, with shining brown eyes, lovely dark hair, and a complexion so petal-*clear*—you'd think Frances' beauty was just happenstance.

But Frances herself says very positively, she *keeps* it that way with her faithful Pond's devotions.

"Skin *needs* regular care," she declares. "I love my daily Pond's Cold-Creamings. They make my skin feel glorious. It *looks* fresher, too."

HOW FRANCES BEAUTY-CARES FOR HER FACE WITH POND'S

First—she *smooths* snowy Pond's Cold Cream all over face and throat, pats it with brisk finger tips to help soften and release dirt and make-up. Tissues off well.

Next—she *rinses* with more luscious-soft Pond's, plying her white-tipped fingers in little spiral whirls around her nose, mouth, cheeks, forehead. Tissues off again. "This double-creaming is *important*," Frances says, "makes skin *extra* clean, *extra* soft. Feels heavenly!"

Use Pond's Frances' way—every morning, every night. Daytime, too, for clean-ups. You'll find it's no accident engaged girls like Frances, noted society beauties, love this soft-smooth beauty care.

Get yourself a *big* jar of Pond's Cold Cream today. You'll like being able to dip the fingers of *both* your hands in the luxurious, big jar.

SHE'S A DARLING! Frances is petite, with wistful brown eyes and skin so baby-soft! "I keep it nice with Pond's Cold Cream," she says. "It's such a grand cream for giving that beyond-a-doubt cleanness and sparkle."

ON HIS FURLOUGHS Paul and Frances are inseparable. While he is away she serves, too—in the Red Cross, at the "Two for One" canteen, and at the Halloran Hospital.

TODAY—more women use Pond's than any other face cream at any price. Ask for this delightful cream at *your* favorite beauty counter.

She uses Pond's

A few of the Pond's Society Beauties

MRS. VICTOR DU PONT, III.
LADY BRIDGID KING-TENISON
MRS. GERALDINE SPRECKELS
MRS. CHARLES MORGAN, JR.
MRS. JAMES J. CABOT

A series of ads for Pond's Cold Cream reinforced the idea of the engagement ring as a prize.

"Diamond Lady" Gladys B. Hannaford worked as a traveling spokesperson for De Beers.

Audrey Hepburn (as Holly Golightly) wore the 128.54-carat Tiffany Diamond in a setting designed by Jean Schlumberger.

The glamorous Elizabeth
Taylor wore her newly
acquired Taylor-Burton
Diamond to the Forty-Second
Academy Awards in 1970.

Taylor couldn't resist this
emerald-and-diamond
necklace from Bulgari.

During World War II, European Jews fled to Palestine (soon to be Israel), where a thriving diamond-cutting industry emerged.

Madame Wellington was the alter ego of Helen Ver Standig, who modeled her Counterfeit Diamonds.

James Buchanan Brady (*left*) loved gemstones so much that he came to be known as "Diamond Jim."

To mark her engagement to Charles, Prince of Wales, Diana Spencer chose a sapphire and diamond ring designed by Garrard.

Australia's Argyle mine produces a range of colored diamonds, from light champagne to cognac.

Rare, vivid pink diamonds are the most valuable stones to consistently come out of Argyle.

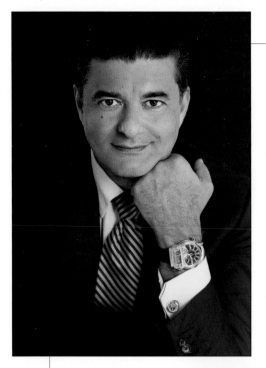

Celebrity jeweler Jacob Arabo
came to be known in the press
as "Jacob the Jeweler."

Fred Leighton
worked with
Forevermark
to create this
diamond suite for
actress Michelle
Williams to wear on
the red carpet.

IF & Co.'s Ben Baller creates unusual custom pieces, like this 47-carat pendant, depicting an LA hat and Air Jordan 11's.

Baller customized this 2014 Rolex Day-Date II President watch with 28.5 carats of high-quality diamonds.

A 2.34-carat radiant-cut yellow diamond, grown in a lab.

In 2014, Pure Grown Diamonds revealed this record-breaking 3.04-carat stone, the largest ever grown in its "aboveground mine."

10

The Meddlers

HOW LITTLE STONES
GOT BIG

New York, Johannesburg, London, and Japan

When Taylor and Burton came home from Italy, they went on a spending spree, as if they were trying to recapture the thrill of their illicit love affair. Burton obsessively chased down extravagant jewelry, spoiling his wife and, to some observers, struggling to overcome the guilt of leaving his wounded ex for her by becoming the most doting husband in the world. In 1968, after four years of marriage, he bought Taylor the Krupp Diamond: a 33.19-carat Asscher-cut stone, previously owned by the German actress Vera Krupp von Bohlen und Halbach, the wife of a munitions mogul who was convicted as a war criminal after World War II. Taylor wore the diamond on her ring finger and enjoyed the irony of owning it as a proud Jewish woman (she had converted in her late twenties after Mike Todd, originally named Avrom Hirsch Goldbogen, died). Burton kept the lavish presents coming. In early 1969 he bought La Peregrina, a perfect natural teardrop-shaped pearl that dated back to the 1500s and had been, at the time of its discovery, the largest pearl ever found.

By then, the country was experiencing an economic slump. The United States was heavily invested in the Vietnam War. Unemployment was up. Experts had only recently started ringing the alarm bells of recession. But that didn't stop Burton, who, in late 1969, saw another opportunity to outdo himself. A stunning 69.42-carat pear-shaped brilliant was going up for auction at Parke-Bernet, the same gallery that had sold him the Krupp and La Peregrina. The public didn't know who was selling it; the owner preferred to stay anonymous and the catalogue had it listed as an unnamed diamond, which offered the new buyer a unique opportunity to christen it to his or her liking. But the Burtons had earned special privileges for being such good customers. A few months before the auction, Parke-Bernet arranged for the jewel, set at the time in a ring with two small side stones, to be sent to Gstaad, Switzerland, where the actors had a home. This gave Taylor a chance to inspect the piece and make sure it was worth the money. Sure enough, the Burtons sent the diamond back to New York along with word to their envoy to bid up to $1 million—more than three times the $306,000 (about $2.1 million today) that Burton had paid for the Krupp.

The gorgeous pear-shaped stone had originally made headlines a few years before when it was found in the Premier mine as a giant 240.80-carat piece of rough. Harry Winston, still the reigning king of large diamonds, bought it from De Beers and then gave it to his cleaver Pastor Colon Jr. to study. After a few weeks spent in intense concentration, Colon was ready to make the first cut. As had become routine at the offices of Harry Winston, the press was invited to Fifth Avenue to witness and document the high-pressure cleaving. At 11:55 on the morning of December 20, 1966, Pastor Colon Jr. took a fateful swing as cameramen lined up to get the shot. A seventy-eight-carat chunk fell to the table. Colon picked it up to inspect it while Harry Winston watched anxiously. "Beautiful," the cutter proclaimed.

Winston exhaled. "Thank you," he said.

Two stones resulted: the seventy-eight-carat offshoot and the original diamond, now slimmed down to 162 carats. That day, Winston estimated that after it was shaped, faceted, and polished, the smaller stone would weigh in at twenty-four carats, while the larger one—the piece that would go on to become the unnamed diamond— would be approximately seventy-five carats. Given the high quality of the rough, he predicted that the two jewels together would bring in $1 million total. This gave the media a particularly tempting lead: the *New York Times* wrote that, for Colon's cut, "one tiny slip can mean a million-dollar loss."

At 69.24 carats, the larger diamond came in slightly underweight and was sold quietly to a private buyer: Harriet Annenberg Ames, one of the seven daughters of Moses "Moe" Annenberg—a wealthy newspaper publisher and rumored mobster—and sister of Walter Annenberg, the American ambassador to the United Kingdom under the Nixon administration. Ames bought the stone for $500,000 (almost $3.7 million today), thinking she would wear it, but realized quickly that she felt self-conscious in a hulking bauble that might make her a target for theft. The cost of keeping it safe in a vault and paying the insurance premium was a prohibitive $30,000 a year ($212,500), according to some sources. Rather than accrue more debt for a luxury item she wasn't even enjoying, Ames contacted Parke-Bernet.

The day of the auction, rumors swirled that the Burtons were interested, and also that shipping tycoon Aristotle Onassis might be bidding for the benefit of his new wife, the former First Lady. There were 162 jewelry lots on offer, and at 1:45 p.m.—the time the sale opened to the public—some eight hundred potential bidders raced beyond the silk cord to find their seats. The catalogue was called "Precious Stone Jewels," and the auctioneer warmed up the crowd with a few laughs. When an item described as a "3/4-inch platinum drill bit" came up to the block, he chuckled. "I guess it's for the man who has everything," he offered, before it sold for $75 ($492).

When it came time to sell the unnamed diamond, members of the audience hushed and straightened. Only a lucky few in the first row could actually see it—even the largest gemstone is still only the size of a paperweight. However, those actually in the running, like the Burtons, had already inspected the jewel. Bidding started at $200,000 ($1.3 million), then quickly jumped to $500,000 ($3.3 million) as the air in the room started to rush. The crowd clapped and cheered as a handful of active bidders—among them diamond dealer Al Yugler, the representative for the Burtons—leapt in front of one another in increments of $25,000 ($165,000). Later, seasoned onlookers would agree that they'd never before attended an auction that felt quite so rowdy and charged. Within minutes, bids had reached $850,000 ($5.5 million), then $1 million ($6.5 million)— the point at which Yugler had been instructed to drop out.

Then the final bid: $1,050,000, or $6.9 million today. It was the highest price ever paid at auction for a diamond, blowing the previous record of $385,000 out of the water. The winner was Robert Kenmore, on behalf of none other than Cartier. The *maison* had anticipated that even the high rollers would back off at the million-dollar mark, and so its strategy was to bid up in the hope that it would be the only one left willing to cross that line. Kenmore worked for the Kenton Corporation, the new majority stakeholder in Cartier after the family cashed out in 1962. Immediately after the sale, the press homed in on gaps in the story. Who was the seller who didn't want to be associated with the unnamed diamond? Kenmore sounded mysterious when he told a reporter for the *Chicago Tribune* that "he was sure the company would have no difficulty selling it." To the *New York Times*, he mentioned obliquely that he was "not a free agent," leading some journalists, like investigative reporter Edward Jay Epstein, who wrote about diamonds in the early 1980s, to suspect a stronger connection between the Burtons and Kenmore than either party admitted.

Sure enough, the very next day it was announced that Richard Burton would be buying the stone after all, but from Cartier, for an undisclosed price that gave Cartier a profit (Taylor later revealed that they paid an additional $50,000 above the high bid). Apparently, Burton had become so distraught after losing the stone that he called the store frantically and arranged for a private sale with unusual terms: Cartier could keep it through mid-November, call it the Cartier Diamond, display it for the public to see in New York, and then ship it to Chicago for the opening of the newest Cartier store at Bonwit Teller. After that, the Burtons could have the diamond and name it whatever they wanted. The parties agreed, leading Edward Jay Epstein to speculate that the whole thing had been a ruse: a backroom deal designed to give Cartier a quick publicity boost before the stone ended up in the hands of the actress to whom it had always secretly been promised.

Certainly, the affair brought attention to Cartier when, on Sat-

urday, October 25, it put the interim Cartier Diamond on display at
the Fifth Avenue branch. By afternoon, a spokesperson for the store
estimated that seven thousand patrons had come through, some who
were willing to wait in line for the privilege of seeing the jewel, and
others who rather ingeniously came in carrying little red boxes—à la
Cartier's packaging—under the pretense of making a return. "Probably there were lots of people who have never been in here before,"
the spokeswoman told the *Hartford Courant*. Five days later, the exhibition remained popular and the syndicated gossip column "Suzy
Says" outed Harriet Annenberg Ames as the diamond's original
owner, while also hinting at the possibility of premeditation behind
the auction's outcome. Whether or not it was true, the whiff of collu-
sion between Cartier and the Burtons was enough to sour critics on
the story. On November 1, the *New York Times* published an unusually
vitriolic editorial: "The peasants have been lining up outside Cartier's this week to gawk at a diamond as big as the Ritz that costs well
over a million dollars. It is destined to hang around the neck of Mrs.
Richard Burton. As somebody said, it would have been nice to wear in
the tumbril on the way to the guillotine."

Indeed, the timing of the sale was imperfect. As executives at De
Beers no doubt celebrated a tale that showed how a diamond could
exponentially rise in value in a matter of years or days, minutes or
seconds, others without any financial stake in the industry wondered if fall 1969 was the right moment for the rich to be flaunting
their money. The same *New York Times* editorial ended with a deep
dig: "In this Age of Vulgarity marked by such minor matters as war
and poverty, it gets harder every day to scale the heights of true vulgarity. But given some loose millions, it can be done—and worse,
be admired." Even the *Hartford Courant*, normally so sympathetic to
the diamond agenda, had this to say about the sale: "Despite the fact

that the Burtons have earned their money, and so are entitled to lavish it where they will, it still seems practically immoral to plunk a staggering sum on so trifling a thing. We live in a world where people are cold, hungry, tattered and ground under by poverty. A million dollars would help quite a lot of them and perhaps even allow them to buy a fake diamond in a dime store to cheer themselves up a bit further."

It's true that the contemporary social and political climate made the world less hospitable to outsize demonstrations of wealth. But perhaps in response—and even more damningly for diamonds—fashion was turning away from the long postwar fascination with Hollywood glamour. At the Parke-Bernet sale, attendees noticed that "members of New York's diamond brigade" were conspicuously absent, and even those who were there to watch echoed Harriet Annenberg Ames's concern that wearing jewels was less of a delight and more of a liability these days. One woman even noted that diamonds didn't go well with the newest styles. De Beers had raised the price of rough another 7.5 percent in 1966, and even before that, the younger generations were having trouble scraping together the money for engagement rings, and were forgoing diamonds for less expensive cultured (meaning man-made, most famously by Japanese company Mikimoto) pearls. In the late '60s and early '70s, cutting-edge designers like David Webb found fresh inspiration in semiprecious stones like jade, prompting headlines like "Bored with Diamonds? Consider Jade" and "Nowadays, a Girl's Best Friend May Indeed Be Jade." Hippies and flower children had no use for Grandmother's rubies and emeralds. In fact, even Elizabeth Taylor's nineteen-year-old son felt the need to distance himself from his mother and stepfather, the unabashed new owners of the Taylor-Burton Diamond.

Michael Wilding told the press that he was moving his wife and infant daughter to live on a commune in Wales, where they planned to grow their own food. "Mother's life seems just as fantastic to me as it must seem to everyone else," he told the *Los Angeles Times*. "I suppose I've always rebelled against it. I really don't want any part of it."

He wasn't the only teenager staunchly rejecting his parents' consumerism. What was a diamond seller to do?

<><><><><><><><><><><><><><><><><><><><><><><>

Since the days when Cecil Rhodes became intent on controlling the entire diamond industry, De Beers had developed quite a few different ways of making sure that the founder's dream remained a reality. The monopoly worked because De Beers effectively dominated every aspect of the diamond industry, from mining operations to distribution to marketing. But maintaining that web of control wasn't easy, and it required the board of directors to respond quickly and strategically every time external forces threatened their power structure and risked throwing it off balance. For instance, De Beers's mines were mostly located in South Africa, but Harry Oppenheimer's Johannesburg was quite a bit more problematic than his father's had been. Apartheid became the law in South Africa in 1948, almost a decade before Sir Ernest died in 1957. But over time, native black Africans were stripped of more and more of their rights, including basic citizenship. In 1961, South Africa was expelled from the British Commonwealth owing to its racist policies; a year later, the United Nations formed a special committee against apartheid. Increasingly, diamond-importing countries were hesitant to support South Africa and its politics. A mix

of moral imperative and professional acumen led Harry Oppenheimer to back anti-apartheid parties in every election, but even
a powerful interest like De Beers couldn't quash the conservative
agenda.

By the early 1960s, there was a sense across De Beers that having
its fate so tightly intertwined with South Africa's was becoming a
real problem. And so it started prospecting for new mines, just in
case the situation in South Africa imploded. De Beers already controlled a large diamond mine in Tanzania (formerly Tanganyika),
and on May 26, 1971, after years of searching in Botswana, it opened
the Orapa mine with a ceremony conducted by the country's president. Orapa was the first diamond mine De Beers ever found on its
own, a traditional open pit in the style of the Big Hole in Kimberley.
Two more African mines followed, both of which posed significant
engineering challenges. Off the coast of Namibia, De Beers discovered diamondiferous soil located deep under the ocean floor. To
effectively collect the stones, De Beers had to do nothing less than
restrain the roaring Atlantic. The process was labor intensive, but
the Namibian yield did not disappoint. Finally, De Beers started
mining in Lesotho, a tiny kingdom surrounded on all sides by
South Africa. There, a kimberlite pipe was discovered deep within
a mountain. The cost of mining at ten thousand feet was enormous,
but the Letseng mine had produced a handful of very large, lucrative stones, including a 601.25-carat one that made headlines in
1967.

By diversifying its geological assets, De Beers defended itself
against the possibility that South African stones would be banned
in any major diamond countries. At any given moment, it might
find it had mined too many diamonds, but it could always stockpile
them and then adjust production to suit global demand. That was

the basic tenet of Cecil Rhodes's strategy, but the principle didn't necessarily account for the mines that might appear in parts of the world where De Beers couldn't get to them first. In those cases, negotiations fell to the syndicate's Central Selling Organization (CSO), based in London, which handled diamond marketing and distribution.

In other words, if—for whatever reason—De Beers didn't have control of a mine's physical operations, the CSO still aggressively courted whichever company or country did, in the interest of buying that company's or country's gemstones and regulating the way in which they entered the market, thereby preserving the larger monopoly. For instance: De Beers entered into this kind of agreement with the Soviet Union in the early 1960s after a pipe was discovered in Siberia. Since just after World War II, when the Cold War began, the USSR had been looking for new ways to ensure a long-term supply of industrial diamonds (industrial diamonds being more important to governments than gemstones, because they're used in the manufacture of modern weapons and are therefore crucial to a country's defense). The Soviets pursued this goal on two fronts. First, the Kremlin sent prospectors hunting for natural diamonds, although they'd never before been found within the borders of the USSR. It also started funding research on lab-grown diamonds, in the hope that a scientific breakthrough would make the country more self-reliant as it became increasingly isolated behind the Iron Curtain.

Ultimately, Moscow was successful on both fronts. Miners struck kimberlite in Siberia after a long period of trial and error in which they learned to overcome the horrific weather conditions: temperatures cold enough to freeze the ground year-round and shatter steel. Eventually, the Siberian open pit became so prolific in its production of gem-grade stones that De Beers couldn't ignore it. The CSO made a deal with the Soviet government to buy all of the USSR's jewelry-appropriate rough. However, analysts hadn't predicted that the Siberian output would be quite so considerable, nor that yield would actually increase over the years, which was very unusual for a tapering kimberlite pipe.

Moreover, the Soviet diamonds, although of decent quality, were almost uniformly small and qualified as melee at less than 0.18 carat. These were not the kind of stones that N. W. Ayer & Son had been promoting; the diamond narrative, honed since 1939, had rather directly implied that bigger was always better, and gifts of larger stones corresponded to more affection. Thanks to De Beers's own advertising campaign, there was little public interest in tiny diamonds. The syndicate had two choices. First, it could end its business relationship with the Soviets. That might solve the melee problem in the short term but it also opened up the very real possibilities that the USSR would either undercut the price of diamonds or find its own way to market small ones and become a serious competitor. This latter scenario was particularly worrisome given the recent upsurge in skilled cutters prepared to handle small-size rough. Back in the old days, the eyes and hands of a master were required to turn melee into beautiful jewels, but that wasn't the case anymore. During the war, Jews in the Nazi-occupied cities of Amsterdam and Antwerp fled Europe and found a safe haven in Palestine, soon to be Israel, where the mayor of Netanya was particularly keen to establish industry for his newest residents. A cutting center sprang up, and unlike the Belgian and American cutters, the Israelis specialized in melee. They innovated the craft by implementing an assembly-line system, wherein each diamond was faceted and polished by multiple people. It made the work faster and also more specific; each worker could focus on one piece of the puzzle rather than be expected to solve the whole thing himself.

This meant that De Beers had no trouble finding cutters for the Siberian diamonds—but neither would the Soviets if left to go it alone. And so De Beers pursued its only other option, which was to continue buying up the Soviet supply of melee—no matter how much came out of the Siberian pipes—and figure out a way to unload it. It was easy enough to sell the stones to traders, wholesalers, and cutters; the CSO had a system in place to make sure that De Beers's best customers bought whatever stones the syndicate was willing to sell. The CSO called it the sightholder system. Every five to six weeks, a select number of prestige buyers who had applied for the privilege were invited to visit 17 Charterhouse Street in London, where they received a small box containing their gemstones for the upcoming season. These meetings, called "sights," were carefully choreographed affairs. A few weeks beforehand, clients—or sightholders, in trade parlance—were encouraged to make requests based on the needs of their businesses, which ranged from factories to high-end jewelers. These requests helped determine the assortment of stones that would be presented to each individual buyer. However, the final decision rested with the CSO and, more specifically, with a gray-haired gentleman named Ernest "Monty" Charles (born Herzog): an old friend of the Oppenheimers' and a former prisoner of war, known throughout the industry for his forbidding gaze.

To his friends, Monty Charles was a convivial chap who loved golfing, gardening, and fine wine, but on the job in London, he was so intimidating that clients were known to fall silent just at the sight of him. The same steely resolve that had helped him survive three years of grueling forced labor in Siam (now Thailand) came across in his business transactions. Charles took into account the clients' needs, but he also paid attention to reports from his team who studied the economic and social conditions of the regions in which the diamonds would ultimately be sold. From there, he put together collections of gemstones that would, ideally, satisfy the buyers while also supporting the general objectives of De Beers. Naturally, the latter was of higher priority. When sightholders arrived, they were escorted into secure rooms and supplied with the proper tools to examine their boxes. After that, they were allowed just two options: to take the entire box as is at a set price, or to pay nothing and leave London empty-handed.

Haggling of any kind was verboten. What this meant in practice was that if Monty Charles wanted to fill a client's box to the brim with melee, he could do just that, and if the client was dissatisfied, his only recourse was to refuse it entirely. More likely, a box would contain enough enticing gemstones—rocks pretty and large enough to be cut and set into one-carat engagement rings, for instance— that the buyer wouldn't feel bilked as he grudgingly took home the less desirable filler stones, too. If a sightholder rejected the offer from Monty Charles, he risked insulting his host and losing the considerable advantage that buying rough diamonds directly from De Beers conferred. Despite the strict provisions, invitations to the sights were coveted precisely because any business that got its stones directly from De Beers was paying the lowest possible markup. A client who refused his box lost his competitive standing in the supply chain and subjected himself to the even less predictable whims of the international marketplace.

That's how an established chain jeweler like Zales, which had been operating since the 1920s and by the 1960s was publicly traded with a number of high-traffic locations, ended up selling unexpected and untested merchandise like diamond cuff links and tie clips. As an N. W. Ayer employee reported, after a meeting with Zales's director of public relations: "He said they got into this business [of selling lower-priced novelty goods] when they first started buying directly from De Beers and had to take a great deal of melee. Faced with the problem of getting rid of it, they came up with these items."

<><><><><><><><><><><><><><><><><><><><><>

After the sightholders left London and scattered across the world, it was their responsibility to figure out how to sell their freshly purchased melee to their own customers. But in this De Beers didn't leave its best buyers totally on their own—after all, it did a good deal of general diamond marketing. Imagine N. W. Ayer & Son's surprise when, at the height of the agency's success with the diamond engagement ring, the client got in touch asking to shift the campaign's emphasis from sparkly large stones to the little flecks of diamond that had long been considered an afterthought. Initially, the advertisers were flummoxed. Internal memos were passed back and forth, debating what, if anything, they should even call melee: "I think calling small diamonds 'miniature diamonds' makes about as much sense and sounds about as up-to-date as calling a child a 'miniature person,'" one member of the team huffed to another. "Furthermore, I think it gives the whole show away." She suggested using melee in diamond picture frames, shoe buckles, dinner bells, and cigarette cases—clearly grasping at straws.

In truth, the challenges associated with melee were manifold. Ayer had to help jewelers come up with ideas for how to use these tiny diamonds while simultaneously creating consumer demand. The team also had to be careful. If they did their job too well, they risked making small diamonds look so precious that they might accidentally undermine the value of the holy solitaire. The answer, they determined, had to do with pinpointing an "under-developed market opportunity." Ayer found what it was looking for in married women: "Only 14% of married women have received a new diamond gift of any kind since their marriages," one internal memo read. "If we make the safe assumption that most married women would enjoy receiving a diamond gift, then we must question the fit of our product (diamond jewelry generally) to this market." A safe assumption, indeed. Diamond gift opportunities included Christmas and birthdays, but also wedding anniversaries, which seemed like exactly the kind of celebration that could be made more festive with a little glitter. Perhaps jewelers might create special pieces to commemorate each milestone—five diamonds for the fifth year of marriage, ten diamonds for the tenth? Eventually, the idea of the eternity ring—a circle of diamonds to symbolize a life full of love— also came about as an ingenious way to get rid of melee. Last, Ayer decided to move the conversation from a diamond's size to its quality, with the suggestion that a finer stone was more desirable than a big one.

That took care of married couples, but what about a new up-and-coming market: the professional women who eschewed marriage and, perhaps in some cases, even bought jewelry for themselves? As the women's movement gained traction, Ayer attempted to target this new customer with a fresh set of ads in glossy magazines like *Vogue* and a clever spin on its time-tested slogan: "A Diamond Is For Now." Maybe the liberated woman didn't want a ring, but would she refuse a classy set of diamond studs from the man of the hour? "You sure know how to stick pins in that old feminine mystique," one 1973 ad read, over a photograph of a beautiful, confident woman wearing a diamond stickpin in her blouse while talking on the phone, making no secret of its intended audience. A 1971 version showed a pair of diamond earrings tumbling from a box of pills, with the copy: "Fast relief. For headache. Tension. Woman blues. Goes to work in seconds. Unlike old-fashioned remedies. An overdose may be a cure-all. A diamond is for now."

◇◇◇◇◇◇◇◇◇◇◇◇◇◇◇◇◇◇◇◇◇◇◇◇◇◇◇◇◇◇◇◇◇◇◇◇◇◇

Even so, try as it might, the syndicate couldn't always perfectly predict how the consumer market would respond to its actions. By the early 1970s, a combination of shifting social values, the new emphasis on melee diamonds, an oil crisis, and an ensuing global recession meant that the very result that De Beers and Ayer had dreaded came to pass: a precipitous drop in engagement ring sales in America. At the height of the 1960s, diamond sellers could count on 75 to 80 percent of couples buying a diamond, but now surveys showed a notable change, not just in buying habits but also in the attitudes of the present wedding generation. Affirmative answers to statements like "Engagement ring is out-dated," "Girl who wants DER [diamond engagement ring] has wrong set of values," and "Diamond is old-fashioned" were climbing upward. De Beers admitted in the *Wall Street Journal* that in 1975 diamond sales had declined by 6.5 percent from the previous year.

After more than thirty years of diamond fever, Americans had officially grown tired of rocks that looked anything like the Taylor-Burton Diamond, not to mention the people who had money to buy them. But De Beers had the foresight to look beyond the United States and adapt the engagement ring campaign for other regions of the world where the tradition was weak or otherwise nonexistent.

The British were already coming around, with over 60 percent of grooms-to-be in 1966 feeling pressure to go to the jewelry store. In the early 1960s, before her retirement, Dorothy Dignam started reading up on the ring's role in French and German marriages, and she reached out to the newly founded Connaissance des Pierres Précieuses (CPP)—France's version of the GIA. For one of its inaugural events, the CPP hosted a gemstone exhibit on an ocean liner and partnered with the Parisian fashion magazine *Marie Claire* to host a six-month diamond contest in which the winner got a three-carat diamond ring and twenty-five runners-up won smaller baubles. Dignam was tickled by the effort, which she described in a memo to her colleagues: "European jewelers are starting a campaign to make the diamond a more significant part of betrothal and marriage as it is in this country. . . . We understand the French diamond merchants have formed a publicity committee and are hoping to get some stories in women's magazines. I'd say we had a slight start over them in that respect!"

Kidding aside, she was happy to pitch in. She arranged to have Ayer's informational booklets translated, and De Beers started advertising in popular magazines overseas. Over the next few years, however, French and German citizens proved difficult to budge. The Germans in particular were happy with their custom of giving a woman a plain gold band, which she wore on her left hand to mark her engagement and switched over to her right hand once she was married. It turned out other cultures had traditions that they singularly cherished. But that didn't stop De Beers from making an unlikely play for Japan: a country that, at least on paper, seemed almost impossible to convert. For one thing, the Japanese had no diamond history; in Japan—unlike Europe and India, where royalty had treasured sparkling gemstones—the elite had never worn them.

Moreover, citizens there almost universally practiced Shinto, a religion indigenous to Japan with a heavy emphasis on ancestor worship and ritual. Tradition dictated that marriages were arranged by the parents of the bride and groom. The Shinto wedding ceremony was time-honored: the bride wore an embroidered white kimono—white being the color of mourning in Japan—to symbolize the young woman leaving her family's house. Instead of exchanging rings at the altar to affirm their commitment, the young couple sipped sake from the same cup.

For its Japanese campaign, De Beers hired the J. Walter Thompson agency, which had more international experience than N. W. Ayer & Son. J. Walter Thompson's strategy was to target the younger generation with ads that emphasized the diamond engagement ring's central role in Western culture, hiring Japanese models with more typically Caucasian features and showing them engaged in activities like bicycle riding or mountain climbing that viewers might associate with a cool and carefree European or American lifestyle. What the advertisers detected among the youth in Japan—and what therefore became the linchpin of their campaign—was a fascination with all things Western. A historically very insular population had been exposed to European and American culture during and after the war, and furthermore, had suffered what was generally regarded as an embarrassing defeat. Particularly among Japanese youth, this created a desire to distance themselves from the older generations and become worldlier. Never mind that the Japanese had no courtship ritual to speak of, or that at least in America courtship had been integral to the success of the diamond as a product. In Japan, the diamond's allure was about more than romance and status—two already quite heady ideals. There, it also represented novelty, and an opportunity to make a break from the war-torn past.

Somehow, against the odds, it worked. Research showed that before the mid-1960s when De Beers kicked off the campaign, only 6 percent of Japanese brides received a diamond engagement ring. By 1981, that number jumped up to 62 percent. Thanks almost entirely to advertising, Japan became the second-largest consumer of gem-grade diamonds in the world, a runner-up only to the United States. Even if sales in America continued to decline and the French and Germans never fully embraced the tradition of giving an engagement diamond, De Beers had successfully bought itself an insurance policy. De Beers had been right about Japan. There, diamonds represented everything that was fresh and new—and the word "hippie" wasn't yet a whisper.

The Inventors

HOW TO MAKE A DIAMOND
(AND PROVOKE DE BEERS)

Paris, New York, and Washington, DC

U nfortunately for diamond industry executives, young, antiestablishment Americans weren't the only threat to their livelihood. Since the earliest days of De Beers, its board had lived in fear of encroachment from a surprising source: the scientific community. The entire monopoly rested on the premise that diamonds were rare earthen treasures. If an alternative source of gemstones was discovered—one that existed well outside the syndicate's reach—then the glittering stronghold built by Cecil Rhodes could potentially collapse.

Making matters worse for De Beers, scientists had been fascinated by diamonds since as far back as the seventeenth century, when Sir Isaac Newton wondered if the hardest substance on earth might actually burst into flames if subjected to high enough temperatures. He was right, but never proved his own hypothesis. However, Newton inspired the subsequent generations of chemists who, through a series of related experiments, eventually came to an unexpected conclusion: the scintillating diamond was just plain old carbon. Graphite was its dull cousin. There was nothing extraordinary about the diamond's chemical composition, so the magic—the special alchemy that separated the world's most precious gemstone from simple pencil lead—occurred in the process of formation.

What exactly that process entailed, nobody knew. The question of how to make a diamond obsessed certain scientists. But it wasn't until 1905 when a young, black-bearded chemist named Henri Lemoine contacted an older man named Sir Julius Wernher that the standoff between science and industry became truly heated. Wernher was a German-born banker who had backed Cecil Rhodes against Barney Barnato during the contentious early years of De Beers. Wernher held interests in many of the country's most prolific mines, including Kimberley, and sat on the corporation's governing board. In his letter to Wernher, Lemoine explained that he had created diamonds in a laboratory that not only were gem quality but also ranged up to one carat in size. Would Wernher be interested in traveling to Paris to witness his process? Wernher found it impossible to ignore Lemoine's offer. If the worst had come to pass—if science had finally managed to replicate the growth that happened deep beneath the *kopjes* in the belly of South Africa—then he wanted to know about it. He made his travel preparations at once.

In Paris, Wernher and a few of his colleagues met the sparkling-eyed scientist at his house on the Rue Lecourbe on the bohemian Left Bank. Lemoine was thirty years old but had a grave air about

him, and commanded respect like a man twice his age. Wernher
was immediately impressed. But then Lemoine launched into his
demonstration and shocked his guests when he began removing
all of his clothing. The spectators questioned his methods, but Lemoine, dead serious but only half dressed, assured them that this
was the only way they'd be utterly convinced by his work. After all,
when the representatives from De Beers saw his flawless product,
they would surely suspect that the diamonds were planted, and
Lemoine wanted to prove that he had no gemstones hidden up his
sleeves or stuffed in his pockets. No, Lemoine assured them as he
started to tinker, there was no trick: this was chemistry at its best.

The De Beers men were entertained by Lemoine's theatrics. In
a perfect world, the silly nude scientist would turn out to be a sham
and they'd enjoy a dinner of sole meunière before returning home
to Johannesburg. But when Lemoine finished his experiment and
produced a smattering of sparkly little rocks, it instantly became
clear that this Parisian magician was going to be a problem. Wernher and his colleagues all examined the product and agreed: these
were definitely diamonds. Immediately, Wernher offered his host
a considerable sum to hand over the recipe. Lemoine declined. He
couldn't put a price on intellectual curiosity. If he'd gotten these
results in such a short time, imagine what he could do next.

The possibilities were endless, and that's exactly what Wernher was afraid of. Eventually, the two men came to an agreement:
Wernher would finance Lemoine to build a proper laboratory outside Paris, and in return, Lemoine would write up his secret formula and deposit it, sealed, at the Union Bank in London. Upon
Lemoine's death, ownership of the document along with all of the
scientist's holdings would revert to Wernher or, if he was already
dead, to his heirs. The arrangement was imperfect, but it was the

best De Beers could do to isolate the threat. Lemoine and Wernher drew up a contract, and then Wernher returned to South Africa, with a promise from Lemoine that he'd be in touch shortly with updates.

The scientist kept his word, and over the next few years, he sent his benefactor photographs of his new and improved accommodations. Repeatedly, Wernher sent money in return. But over time, he began questioning his own judgment. Maybe his fear of the unknown had caused him to act rashly. If Lemoine was hard at work, what exactly was he producing? There were rumors coming out of Paris that Lemoine was living the high life and had even bought himself a flashy new motorcar and a mansion down the street from the Moulin Rouge. More troublingly, Wernher and his partners heard other whispers that Lemoine's diamonds looked just like Kimberley stones because that's exactly what they were—stones purchased from a French dealer.

Wernher confronted Lemoine, and when the scientist couldn't answer his concerns to his satisfaction, Wernher had him arrested for swindling him out of $320,000 ($7.9 million today). The press embraced the story, which was David and Goliath with a twist, the great behemoth of De Beers outsmarted, the everyman a wily trickster—if, of course, the accusations were true, and he was in fact defrauding his patron. The French were divided on that point. Everyone agreed that Lemoine's diamonds were real, but whether or not they'd been surreptitiously planted was a question that came between friends and neighbors. The case went before a judge in the winter of 1908, and the public followed the proceedings, captivated. The story became *l'affaire des diamants*—the affair of the diamonds—in the papers. A young writer named Marcel Proust published an experimental novella called *The Lemoine Affair*. When,

in late January, there was a horse race at the Hippodrome de Vin-
cennes with a horse called Diamant ridden, coincidentally, by a
jockey named Lemoine, it gave citizens the opportunity to bet in
line with their beliefs. The race resolved faster than the trial, with
Diamant an unlikely winner and his supporters laughing all the
way to the bank.

And speaking of banks, Wernher's lawyer suggested that the
whole suit could be easily resolved if only Lemoine would unveil
his secret recipe and present it before the court. Lemoine vehemently objected, arguing that he and Wernher had an agreement
and, moreover, he was sure Wernher was using the legal system
to violate it by prying the formula from him before his death. As
the trial unfolded, the court heard testimony from various witnesses who, like everyone else, seemed to have very strong opinions about Lemoine. Leading scientists, who had been invited to
view the Frenchman's experiments, defended him. Experts from
the diamond industry were equally biased. One particularly compelling witness testified that he had met Lemoine years before in
Paris when Lemoine first became interested in the manufacture of
artificial diamonds. According to this dealer, the defendant's motives had nothing to do with scientific inquiry. Rather, he wanted
to weaken consumer confidence in De Beers by spreading word
of his perfect fakes. The value of De Beers stock would drop, and
then Lemoine—in a calculated move worthy of Cecil Rhodes—would
buy up so many reduced-priced shares that, eventually, he himself
would become part owner of the syndicate.

Lemoine was held in custody for months while the issue of
whether the French judiciary could compel an English bank to
relinquish the secret formula played out. Finally, in April 1908,
the English court—which until then had steadfastly protected Le-

moine's deposit—reversed its stance and conceded to the French legal system. With the sealed envelope on its way to Paris, Lemoine threw himself on the mercy of the French court. Did the judge not understand the terrible loss Lemoine would incur if the recipe were made public? Its value would plummet. Lemoine's hard work would have been for nothing, and Wernher would have no further incentive to honor their contract.

The judge recognized his complaint and agreed to an alternative test. If the scientist could perform his experiment before the court and make diamonds, then that would be enough. More time passed while Lemoine supervised the construction of a new electric furnace—one that he'd never actually touch, to prevent the possibility that he might sprinkle it in advance with mined diamonds. Ten new crucibles—the containers in which the stones were supposedly grown—were imported from Germany, sealed. On the day of reckoning, one would be chosen at random, and Lemoine would instruct a representative of the court in how to mix the elemental batter and bake it to a glittering white. "Fraud will be impossible," the papers said.

Then came the fateful June morning, and Lemoine, owing to unforeseen circumstances, asked for a deferment, which the judge granted. June 17—just shy of two weeks away—was his last chance. That day, more than six months after Lemoine's arrest, there would finally be an ending to the story that had entranced the interested public like the best Dickens novel. " 'L'affaire des diamants' . . . [made] the famous Dreyfus case seem flat, dull, and commonplace," one journalist wrote. Lemoine's supporters held out for his inevitable vindication. Others couldn't help hoping for the triumph of the charlatan, guilty in the eyes of the legal system but nonetheless victorious for having kept up the ruse for so long.

And then, on the morning of Lemoine's court date, the judge waited, and waited, for the man to appear. He never did. A last twist in the tale: Henri Lemoine had fled Paris.

When the judge opened the envelope, Wernher's worst suspicions were confirmed. Lemoine had nothing. The formula was nonsense, and good only for extorting the hundreds of thousands of dollars that had lined a swindler's pockets. Wernher never got his money back. But on the upside, he and his colleagues at De Beers got to rest a little easier knowing that still no one had successfully synthesized diamonds. It would stay that way until 1955.

∞∞∞∞∞∞∞∞∞∞∞∞∞∞∞∞∞∞∞∞∞∞∞∞∞∞∞∞

After Lemoine, other scientists tried for a breakthrough. Percy Bridgman, a Nobel Prize–winning Harvard physicist, made progress but never made diamonds. However, as he told *Scientific American* in 1955, the fact that his experiments didn't yield an ice-white gemstone might just have been a blessing in disguise: "I have often encountered the belief that the successful solver of this problem would be in danger of his life from the Diamond Syndicate."

It was an exaggeration, but as the Lemoine case so many years before had proved, anyone who grew diamonds could expect to attract a good amount of attention from De Beers. Yet in 1953, a lab operated by ASEA, Sweden's preeminent electric company, managed to create the first artificial industrial diamonds without making waves at all—in fact, the team was so secretive that a few years later, when a competing American lab—General Electric— pulled off the same trick, the Americans got all the credit for being first. Executives at General Electric had practical motives for pursuing a proprietary formula for industrial stones. General Electric used diamonds to shape the tungsten in lightbulbs, and no other material did the job as well, which meant that the company, like the American government in wartime, was beholden to De Beers. For as long as De Beers controlled the vast majority of the world's mines, this dependence left General Electric vulnerable. In 1950, it put together a team and launched the top-secret Project Superpressure.

General Electric hired a crew of PhDs and moved them from their various posts to Schenectady, an industrial town north of Albany that fell into a deep freeze every winter. The original Project Superpressure team was a tight-knit set of four, and when a Mormon scientist from Utah named H. Tracy Hall joined the group, he felt excluded by his colleagues. Still, they worked together on the assignment's most vexing questions, like how to mimic the unique conditions present in a kimberlite pipe. Four years into the team's tenure at General Electric, they achieved their first real triumph. There was mounting competition among members, as the majority went one way but Tracy Hall, the outlier, pursued his own instincts regarding the design of the diamond-growing apparatus.

On December 8, 1954, Herb Strong, one of the original foursome, ran Experiment 151, which lasted sixteen hours. The next morning, he was disappointed to find that nothing but a sad mass of melted iron—one ingredient in his complex recipe—had appeared in the chamber. He sent the iron to the metallurgy lab for testing to see if there was anything to learn from yet another flop. Almost a week later, a tech alerted him to the existence of an unusual crystal hidden within the sample. Not only did it appear to have emerged from the thick metal, but it was also incredibly durable, almost too hard to polish. Strong was dumbfounded, and he and the group initiated a round of testing to determine whether or not it was a diamond.

Meanwhile, Tracy Hall rushed to run his own follow-up experiment using the method he'd pioneered. With the help of his colleagues in the workshop, he'd built his machine in the off-hours, acting on a hunch that the other scientists had discouraged. Now, armed with Strong's winning recipe, he readied his machine, which had yet to produce a diamond. He waited just thirty-eight minutes for the elements to cook. On December 16, Tracy Hall saw something glittering in the darkness. The experiment worked. For him, this accomplishment represented more than the culmination of years of scientific inquiry; it was something else, too—a tantalizing mix of victory and vindication.

That feeling was reinforced when, in the following weeks, Strong was unable to duplicate his results while Hall, on the other hand, made diamonds many times over. Like Strong's crystal, Hall's stones were industrial-grade: no one would think to mount one in a Tiffany Setting. But fine jewelry was never the goal, and what General Electric had set out to do, four years later Project Superpressure delivered. On February 15, 1955, the project members went public with their story by way of a press conference in Schenectady. As the *Baltimore Sun* reported in an otherwise bullish article: "So far, [the GE lab] has made only $10 worth of diamonds, mostly in the form of a fine dust. That $10 retail value includes the first 'big' diamonds—only one sixteenth of an inch long—which cost hundreds of thousands of dollars if all expenses of four years of research are included." However, the overarching message was positive: this technology was revolutionary. Apparently, investors agreed. The day of the announcement, shares of General Electric on the New York Stock Exchange showed unusually high activity and by late morning had risen 4.125 points over the previous day's close. At the same time, De Beers Consolidated Mines dropped by almost 3 percent.

Next, General Electric applied for a number of US patents to protect its methods, and soon afterward, the corporation was surprised to learn that the government had a similar goal in mind, although it recommended an entirely different strategy. Rather than supply the patents, which would make both the chemical formula and the blueprints for Hall's apparatus public, the administration wanted to keep the applications under wraps, to ensure that America would have its own domestic industrial diamond supplier. The patents weren't granted until 1959, and during the interval the Project Superpressure team—minus Tracy Hall, who left General Electric soon after the press conference because he felt he hadn't been appropriately recognized for his individual work—was able to fine-tune the process. By the time its artificial diamond grit went on sale, the team was confident about its product. Not only had the members accomplished the impossible by synthesizing diamonds, but in the years since the announcement, they'd even created a diamond alternative, which they trademarked as Borazon: a material so strong it actually scratched diamond, previously the hardest substance known to man.

Both inventions generated enormous commercial success. Then, in the summer of 1959, the US government finally declassified General Electric's research, and the scientists were free to publish it. They did just that, and a few months later, in September, the company applied for international patents. GE had waited long enough, and had no intention of being blindsided by some other up-and-coming laboratory. Just a few days afterward, GE executives were undoubtedly shaking their heads at the thought of what might have happened if they'd held out just one week longer. Another patent application had been filed, this one from halfway around the world, in South Africa.

De Beers had made its own synthetic diamonds.

De Beers presumably turned its focus to synthetics for the very same reason it agreed to keep buying up the stores of Soviet melee: its business model required it to dominate every corner of the market. However, the timing was not in De Beers's favor. Given that General Electric had originally filed a US patent application in 1955, there was no use contesting GE's claim in the States. However, with just a few days separating their international applications, the De Beers leaders made a decision to challenge General Electric's advantage. They filed numerous lawsuits, and the battle went to court in South Africa, where, just as in the Lemoine case, it took on a life of its own. The proceedings began in 1959, and they dragged on for the next six years.

⬦⬦⬦⬦⬦⬦⬦⬦⬦⬦⬦⬦⬦⬦⬦⬦⬦⬦⬦⬦⬦⬦⬦⬦⬦

The manufacture of synthetic industrial diamonds represented an impressive advance for the scientific community, as well as a boon for governments and factory owners who stood to benefit financially from some long-awaited competition in the marketplace. But for the men who bought gifts of romantic diamonds and the women who loved to wear them, this accomplishment meant almost nothing. Readily available bort was great for the country, but what about the woman who loved diamonds but couldn't easily afford them? Or the lady who liked to dress up her outfit with a little sparkle but who wasn't willing to risk losing a small fortune every time she walked outside?

For those customers, there was Madame Wellington: the doyenne of counterfeit diamonds. Madame Wellington was the alter ego of Helen Ver Standig, a brassy, chain-smoking, gravel-throated

Washington, DC–based entrepreneur who stood only four foot eleven and made her fame selling outrageous imitation jewels. Late in her life—after she'd spent decades running the business called Wellington Jewels, and was so well known in the capital that she and Katharine Graham of the *Washington Post* were the city's two most prominent women—she arrived at a party thrown by a friend who had just recently remarried. His new wife was a much younger Vegas showgirl with a tendency to overindulge. Good and soused, she approached Ver Standig and complimented her on her jewelry. "Helen dear, what a lovely ring you have on, it's a shame it's not real," she teased, assuming Ver Standig was modeling one of her own products.

Ver Standig, who liked her liquor, too, and who was known for sitting down with men twice her size and drinking them under the table, got up and ambled toward a plate glass window. She slowly ran her ring down the length of the pane, leaving a long scratch.

"Whoops!" she said as her hostess gaped. "It was one of the real ones. So sorry."

That kind of irreverence was typical of Helen Ver Standig. She was the daughter of an Austrian couple who left Europe for America because her mother had been previously married and the divorce left her an outcast. The Van Stondegs escaped to a poor neighborhood in Washington, DC, and Helen's father opened a small tailoring business. Helen worked at the counter every day after school, and one afternoon a handsome young man walked through the door. Maurice "Mac" Belmont VerStandig,* a Jewish newspaperman from Boston, was traveling up and down the Eastern Seaboard, looking

. . . .

* Helen spelled "Ver Standig" as two words, but the rest of her family uses a one-word spelling.

for distant relatives who might have emigrated from Europe during World War II. Helen was only a teenager, and the initial attraction between them was so potent that they decided to get married six days after their first date (they did not turn out to be long-lost cousins). "We couldn't wait to jump in the sack," Ver Standig once famously told a reporter for the *Washington Post*.

After that, Helen's career as a shopgirl was over. She and Mac bought newspapers in Cranston, Rhode Island, and Greer, South Carolina. In Greer, they had trouble courting local business owners because the residents suspected that the VerStandigs weren't actually committed to staying in town. They assumed—perhaps rightly—that the couple planned to live in Greer for a year or two and then sell the paper, and they made it clear that they preferred to advertise with people who were from the area or, at the very least, intended to put down roots. The VerStandigs got the message, and Helen, who had just finished high school, came up with a creative solution. She went to the local cemetery and bought two side-by-side plots. Then, she figured, the locals would believe that they intended to stay put for the long haul. It worked.

Eventually, however, the couple returned to her hometown, Washington, DC, and had two children. They moved on from newspapers and started an advertising firm called M. Belmont VerStandig Inc., landing high-profile clients like Geico, Marriott, and the US Army. The VerStandigs were a charismatic duo who attracted all sorts of talented and powerful people. Once, at a football game at Griffith Stadium, when her son and daughter were both quite young, Helen handed her neighbor a twenty-dollar bill and instructed him to go buy her children hot dogs and hot chocolate. The stranger balked—he was there to watch the game, not take orders—but was impressed by the woman's moxie. That man turned out to be Joe Nesline, the gangster boss of Washington, DC. He and Helen became lifelong friends. Years later, when Nesline was wanted for murder, the VerStandigs had no qualms about harboring him in their basement. John, their son, remembers coming home from elementary school to see a line of police cars up and down the block. He supposed that Uncle Joe had finally been caught. But as he approached the house, he slowly began to process the real reason for the cavalcade: his father was having coffee with Richard Nixon on the front porch. Unbeknownst to the vice president, his mother was quietly playing cards with a wanted criminal just beneath them.

By the mid-1960s, the VerStandigs once again were in search of a fresh challenge. Mac had his eye on the hotel industry while Helen was more interested in real estate, but inspiration came from somewhere entirely unexpected: a solo boat trip Helen took to Europe. On the way home, she met and befriended a professor from Switzerland, who talked to her about imitation diamonds. He had a connection to a lab that was experimenting with strontium titanate, a compound that looked and behaved very similarly to diamond, especially when it came to dispersing light. He took down her address and promised he'd send her some sample stones. Back in DC, Helen sat down with Mac at the dinner table and told him what she'd learned. She was excited, and he also saw the potential in selling good-quality simulated (or "simulant") diamonds. When the VerStandigs received an envelope in the mail, postmarked Switzerland and filled with lovely cut and polished gemstones, they were ready to take a gamble. Their son, John, explains, "Now understand, this would be like Kenny Feld, who owns Ringling Bros. and Barnum & Bailey Circus, stumbling across a unicorn. They weren't dumb. They realized this had great promotional value."

The VerStandigs had no background in jewelry, and so they trod carefully, making the decision to test the product with a small direct mail marketing campaign. To gauge public interest, they took out ads in a few second-tier newspapers offering stones at forty dollars a carat (around $300 today). Serendipitously, it was a moment at which De Beers was ratcheting up its prices. Within days, the VerStandigs had a 4.5 percent response rate, more than double the number that would have qualified as success in their minds.

It was all the information they needed to feel comfortable moving forward. They purchased the German supply source, sold the advertising agency, and started brainstorming names for the product. "Afrique diamond" was a front-runner, as was "Mona Lisa": perhaps a nod to the enigmatic look on a woman's face after she received a compliment on her faux jewels. But ultimately, they landed on Wellington, which sounded established and trustworthy. Helen, whom Mac called Madame, soon became the company's spokesmodel, and the VerStandigs commissioned their friend Al Hirschfeld, the *New York Times* caricaturist, to draw her. The resulting portrait captured her brilliantly: a wink, a tangle of jewelry, a towering tiara, a sardonic, toothy smile. Madame Wellington was Queen Elizabeth crossed with a bawdy hostess, the perfect combination for a business that trafficked in imitation luxuries.

The couple set their sights on bigger outlets: the *San Francisco Examiner*, the *New York Times*, the *Wall Street Journal*, the *Chicago Tribune*. They drummed up cheeky copy to go along with the sketch. The use of the word "counterfeit" in the product name was as unexpected and bold as the woman behind it. What made Madame Wellington different—aside from the consistently good quality of her jewelry—was that she held up her product's lack of authenticity as its best feature. Diamond substitutes had existed since the days of the diamond horseshoe; Sears, Roebuck & Co. even sold cheap

phonies called Crystaline diamonds in its catalogues at the turn
of the century. Other distributors advertised "Golconda Gems"
and "Genuine Parisian Diamonds," which were inexpensive white
stones with intentionally confusing brand names, and in general,
those who wrote the ads for imitation diamonds always seemed to
let their sense of inferiority creep in. The VerStandigs, on the other
hand, hit their customers with the truth. One 1975 ad that ran in
the *New York Times* was written in Madame Wellington's voice: "I'll
be the first to admit that a gorgeous, genuine diamond is—or once
was—'a girl's best friend.' But today, when you have to pay at least
$1,000 for a fine quality, one-carat solitaire, who needs that kind
of friend?" The signature lines on the same ad were unprecedent-
edly frank about the pros and cons of a simulant, all the while sus-
taining a saucy tone: "Wellington Counterfeits have virtually the
same brilliance as perfect real diamonds, and even more 'fire' . . .
Wellington Counterfeits are not as hard as diamonds, so if glass-
cutting is your thing, stick to the Real McCoy."

The VerStandigs encouraged their customers to ask themselves:
What kind of sucker would pay money for a real diamond? Their
frauds shone as brightly as Jager stones. The woman who wore Wellingtons would shine even brighter, knowing she paid next to nothing for sham jewelry that everyone around her mistook for genuine.

The approach worked better than the VerStandigs could have
hoped. The key to their success wasn't just in their ad copy, but also
in the way Wellington handled its faux diamonds. Yes, they were
fake, but everything else about the jewelry—the cuts, the settings,
the precious metals—was completely authentic. With stones set in
fourteen- or eighteen-karat gold and a team of designers who could
mimic the artistry of high-class houses like Harry Winston or Van
Cleef & Arpels, Wellington offered its clients an accessible taste of

extravagance. Helen Ver Standig liked to tease the press with tantalizing blind items that underscored the racy connotations of a
sham diamond: "Honey, I'm responsible for more cheap weekends
than any madam in the country," she was known to say. After they
were purchased, she didn't care how the stones were used. If a man
wanted to lie by omission and let his lover believe he'd sprung for
the real thing, so be it. If a small country needed its crown jewels
updated but couldn't afford Cartier, then its princess would make
appearances in a Wellington tiara and Helen Ver Standig would
take her secret to the grave.

Wellington Jewels was conceived as a mail-order company, but
within a few years, it expanded to brick-and-mortar stores. Once
again, the VerStandigs kept their expectations measured. The
problem with the direct mail business was that after Christmas,
no matter the product, a company could expect roughly 10 percent
back in returns. This meant that every January, the Wellington
office was glutted with excess inventory. The VerStandigs came up
with the idea to open a storefront in DC to handle that merchandise.
They found a modest spot on Connecticut Avenue and prepared for
the grand opening. The first day they unlocked their doors, there
was a line around the block and the Wellingtons on hand sold out.
Soon, the company had storefronts in Tysons Corner, Virginia, and
Chevy Chase, Maryland. Helen found the DC flagship a reliable
manager: Joe Nesline's wife.

Mac passed away in 1972, and the widowed Helen Ver Standig—
who had become almost indistinguishable from her alter ego—
continued building her estate. She franchised Wellington Jewels,
and by 1981, it was an over-$20-million-a-year enterprise ($54 million today). Its fearless leader used some of the profits to invest in
radio stations and "passion projects," including her commitment to

the causes of gay rights and AIDS research. Ver Standig's entrepreneurship earned her a standing teaching gig at the Wharton School of business, and when she wasn't on the clock, she spent her free time traveling the world. With the help of her two grown children, she was able to keep the business entirely family-owned—she entertained buyout offers, but for years and years she declined them, before eventually selling to QVC in 1992.

"My customers love it when they are being robbed," Helen Ver Standig told *People* magazine in 1981. She encouraged her patrons to think of Wellington Counterfeit Diamonds as nothing less than their own private jokes on the world. The VerStandigs proved that if given the opportunity to buy convincing, affordable sham diamonds, many people would take it—no matter what they told their friends. "Everybody has peacock syndrome," Ver Standig explained. "They want to look wealthy but without the risk and the money. The only people who won't buy fakes are the hookers—they want the real thing."

◇◇◇◇◇◇◇◇◇◇◇◇◇◇◇◇◇◇◇◇◇◇◇◇◇◇◇◇◇◇◇◇◇◇◇◇

The physical differences between Wellington Counterfeits and mined stones weren't necessarily obvious to the average admirer, but any gemologist worth his or her salt could distinguish the phonies from real diamonds. This allowed for a tidy hierarchy: synthetics had their benefits, but they were not as dear—economically but also emotionally and intellectually—as the gems that came out of the ground. Whether or not a buyer cared was another story, but no one would argue that in terms of intrinsic value, the true diamond remained unrivaled. In this way, no matter how popular Wellington Jewels became, they ultimately posed little threat to the diamond narrative. Quite the contrary: the VerStandigs arguably reinforced it by showing how intensely Americans wanted to be seen wearing luminous white jewels.

However, lab-grown gem diamonds—stones that were visually and chemically indistinguishable from mined ones—were much trickier. After years of back-and-forth in the South African courts, General Electric won the lawsuit against De Beers, with De Beers ordered to pay GE an undisclosed sum that was rumored to be somewhere between $8 million and $25 million, plus royalties. Feeling vindicated, the remaining members of Project Superpressure turned their attention to the question that still preoccupied chemists: Could jewelry-grade diamonds be created rather than just discovered? If so, the power dynamic in the industry, for the first time since before the turn of the twentieth century, had the potential to turn away from a certain multilimbed corporation with the resources to dig into the bowels of the earth.

In May 1970, the scientific world got its answer. General Electric succeeded in growing beautiful colorless diamonds—slowly but surely, at a pace of about one carat per week. Rush the process, and the samples came out discolored or otherwise marred. It was prohibitively expensive, and when the team members announced their latest triumph to the press, they made sure to point out that they "[didn't] know whether it will ever become possible, in the future, for these G.E. diamonds to compete economically in the gem market. The purpose of this press conference is to announce a scientific achievement, not a product." The following day, De Beers responded, keen to let readers know that if GE's work caused any tremors, they weren't originating in Johannesburg or London: "We . . . congratulate them on that achievement. It does not alter De Beers' plans for the future."

12

The Illusionists

HOW THE MAGICIANS PROTECTED THEIR TRICKS

United States and Europe

A fter the team at General Electric grew their first jewelry-grade diamond, the industry's collective anxiety about science seemed to diminish. A lab that produced a carat's worth of rough per week was hardly in a position to challenge the most prolific diamond mining company in the world. But just a few years later, another threat started to build, this one from an even more unexpected place: the buyers themselves, typical private collectors. By the late 1970s, Americans had grudgingly grown accustomed to volatile economic conditions: a recession, huge fluctuations in the stock market, and vast unemployment. Abroad, the value of the dollar was unstable. Normally, this kind of environment would be inhospitable to the diamond industry, given that luxury items are the first thing buyers give up when times get tough. But this time, something was different, and as the 1980s approached, jewelers began to notice an unusual trend among their clients. Consumers had always been curious about whether or not their purchases would appreciate, but that line of questioning typically followed well behind discussions of the stone's aesthetic and romantic qualities. Now, suddenly people were inquiring about "investment diamonds." They had read about them and heard talk of them on the news. They wanted to buy gemstones the way investors bought low-risk stocks and bonds, with the intention of keeping their money safe for a spell before eventually selling high.

The assumption was that the value of a good-quality diamond always increased, like the works of a great artist. Investors who had been burned by the stock market turned to diamonds as a charmingly old-world commodity. Unlike precious metals, diamonds and other gemstones are not speculative items pegged to international financial exchanges. Something about them felt reliable, crash-proof: quite literally rock solid.

But the chatter around investment diamonds was inflected with an undertone of paranoia, too. There was a sense that in the event of financial collapse, entire underground economies could run on diamonds, the same way Jewish refugees had been able to smuggle and barter their tiniest, most portable assets during the dark years of World War II. Of course, the America of Nixon and Ford was nothing like Hitler's Germany. But for those who felt uncertain, the diamond's strength was a source of comfort. Many diamond sellers saw the trend as a lucky twist of fate, yet another way to market gemstones at a moment when jewelry sales could have taken a devastating turn.

However, there was a problem with this logic, which prioritized immediate profits and didn't take into account the depletion in value that would occur if everyone who invested turned around and sold at once. The very idea of investment diamonds was flawed because it relied on assumptions that actually ran in direct opposition to certain truths about how diamond prices were calculated. This new, fleetingly lucrative trend could cause serious trouble for the diamond world, and only a handful of industry people were willing to say it out loud.

∞∞∞∞∞∞∞∞∞∞∞∞∞∞∞∞∞∞∞∞∞∞∞∞∞∞∞∞∞∞∞

One of those loudmouths was Helen Ver Standig, who took her qualms about investment diamonds to the press. She occupied a unique position because while she personally stood to gain if consumers let go of the notion that ownership of real diamonds had long-term advantages, she also benefited from the popularity of mined gemstones and would have a harder time selling her simulants if the mystique around diamonds fell away entirely. Given that both things were true, there was nothing to stop her, already a bit of a firebrand, from speaking her mind. In a long 1978 profile in the *Washington Post*, she explained why a diamond was not, in fact, like a Picasso: "If you own a Picasso, he's dead, it's the end of his productivity and that's an investment. But there is no shortage of diamonds, and De Beers sells whatever it wants. Just try and sell back a diamond." She specified what might happen a few years later in the *Baltimore Sun*: "Look at it this way: when the jeweler sells you a stone, he's making a profit. If you come back to him several years later to resell him that same stone, he still has to make a profit. So, he doesn't buy it at the same price he sold it to you. He buys it at probably half the price you paid him. That's good business."

Ver Standig's insinuation was twofold: The investor, who realized that he'd have to take a loss on his diamonds, would be furious and sour on gemstones. But also, it was bad business for the jeweler to sell a product by touting its probable appreciation, all the while knowing that even if a diamond's retail value increased, the owner would still be forced to sell at closer to wholesale prices. But again, for many jewelers, a bird in the hand was worth two in the bush. Small, local storefronts reported that business was booming. Others, like Madame Wellington, warned of the possible short-sightedness—or, worse, full-scale opportunism—of jewelers who jumped on the bandwagon.

From within the industry, the conversation felt deadlocked; there were persuasive arguments coming from both sides. Reports from the press were generally bullish, and in September 1978, *Newsweek* published a feature presaging "The Diamond Boom," which emboldened investors. Meanwhile, in New York, an established investigative journalist named Edward Jay Epstein felt himself drawn to the conversation about diamonds. He had no experience writing about them, and in fact his previous work couldn't have been further afield: as a young reporter, he'd made his name writing about Cold War spies. However, years later he sought a change, not least because the tenor of the Cold War was shifting. He pitched a few ideas to the magazine *GEO*, a mostly photographic German publication that had recently expanded to the United States. The one that caught the editor's attention was about African diamond mines.

De Beers was one of *GEO*'s major advertisers. Both magazine and syndicate loved the idea, and they gave Epstein carte blanche.

Harry Oppenheimer was so pleased to show off his dynasty that he flew Epstein around Africa in the company's private planes, arranging for him to be escorted from South Africa to Namibia to Lesotho. As a journalist, Epstein saw impressive, expertly run mining operations, with the grunt work done almost entirely by machines—a far cry from the dangerous conditions that were par for the course in the days of Cecil Rhodes. He also flew with De Beers executives who let down their guard with him, and he ate meals with sales managers and engineers who were happy to share their stories over a few rounds of drinks. The plan was for Epstein to do his reporting, and then a photographer from *GEO* would follow behind and take pictures of the mines for publication. But over the course of his trip, Epstein began to suspect there was more to the De Beers outfit than just prospecting and outsmarting the earth. "I realized there was a whole cartel controlling the world of diamonds," Epstein recalls. Suddenly, the man who spent years trying to expose the activities of Cold War spies found himself entrenched in another surprisingly complex network. As it turned out, diamonds and the secret world of underhanded politics weren't so different after all.

When Epstein returned home, he started researching the history of De Beers, and his tenacity soon led him to N. W. Ayer & Son. From there, he pieced together a ten-thousand-word story for the magazine about the diamond syndicate, beginning with Cecil Rhodes and outlining the many ways that De Beers had systematically fashioned a myth around diamonds: he called it "the diamond invention." Epstein remembers: "What [De Beers] succeeded in doing, which is one of the most brilliant things I know of in any industry . . . was to put it in the mind that [the diamond had a] function and it was a way to get the girl, and to prove to the girl that everything was serious. And then the size of the diamond, which proved to the woman's family that the man was worthy. They did it!" They did it—by advertising, but also by stockpiling and mercilessly ousting their competitors. Epstein turned in his article, which said just that, to his editor at *GEO*.

The editors killed the piece. De Beers was one of *GEO*'s best advertisers. And so Epstein took his ten-thousand-word article and kept working on it until it came in around two hundred pages. *The Rise and Fall of Diamonds: The Shattering of a Brilliant Illusion* was scheduled to be published by Simon & Schuster in May 1982, and Epstein argued an interesting and timely theory. So far, De Beers had managed to overcome any threats to its power by using a mix of shrewdness and intimidation. That approach had served the syndicate well. But now there was a problem so insidious that it had the potential to take down the entire diamond market, and not even a worthy giant like De Beers could stop it.

That problem was investment diamonds. Epstein found the whole trend dangerous for a few reasons, many of which overlapped with the ones enumerated by Helen Ver Standig. But to him, one of the biggest issues had to do with private citizens keeping small stockpiles—in effect, slowly but surely wresting control from De Beers. If enough people kept their own stores of diamonds, and then paid close attention to global prices in order to make sure that the value of their investment would never fall—well, it was a recipe for disaster. It's disastrous because any nervous investor who saw a slight decline in diamond prices would scramble to sell immediately. If enough people did the same thing, it would mean that private owners en masse were suddenly flooding the marketplace. This would cause values to plummet—that dreaded imbalance of supply and demand—and no one would be able to get his money back. Ultimately, in Epstein's view, there was only one possible outcome: confidence in diamonds as a product would be fatally undermined.

This was the worst-case scenario, which, given his reporting, Epstein regarded as inevitable. It also became the focus of an article adapted from his book, which ran in *The Atlantic* on February 1, 1982, and was called "Have You Ever Tried to Sell a Diamond?" It exposed the De Beers monopoly and the details of the N. W. Ayer campaigns, and got so much attention that De Beers pulled advertising from the magazine. The article remains one of the most popular and oft-cited pieces of diamond journalism to this day. Epstein remembers that when it first published, people were "shocked" by the material. By exposing the truth—diamond prices were artificially inflated—he challenged a symbol that had become, among other things, part of the firmament of American courtship. This was even more sensitive than the suggestion that diamonds were as unpredictable as the stock market.

De Beers was furious. It had hosted Epstein thinking he was writing a photo-illustrated piece about the beauty of the diamond regions and the efficiency and safety of modern mines. "What they said to me," Epstein remembers, "is more or less this: Your book would be fine if it were only read by diamond dealers or people in the diamond business because it shows how valuable we are, there would be no diamond business without us, we are it. If you killed De Beers, diamonds would drop in value, we'd all be out of business. We like that part of it. But then again, if the public buys your thesis they won't buy diamonds."

But as Epstein tells it, the public didn't "buy" his thesis—at least, not exactly: "I remember a good friend of mine . . . he said, your diamond book changed my life, it was so great, I have to buy a diamond for my wife, it's our anniversary, what should I do? And I said, oh, you can buy a diamond really inexpensively by going to this auction. He said, are you kidding, my wife would kill me. The whole point of diamonds is to spend a lot of money." In other words, even if readers absorbed, intellectually, the point that diamonds were a manufactured symbol of love and they were surprised, or even disappointed, to hear of it, it didn't take away from the fact that the act of giving a romantic diamond had been ritualized. For generations, the diamond had filled a gift-giving void between lovers, just as De Beers had intended. As Epstein learned, it takes more than mere background information to disarm a loaded gesture of love. To most people, the fact that someone once set out to market diamonds doesn't make the act of giving or receiving one any less special.

As for investment diamonds, "Have You Ever Tried to Sell a Diamond?" and then the subsequent book did have some effect, particularly with exposing dishonest opportunists who were cold-calling homes and then offering sealed packets of diamonds to the gullible folks at the other end of the line. Representing companies with official-sounding names like De Beers Diamond Investments Ltd. (no affiliation), the callers assured anyone who stayed on the phone that the stones they were selling were guaranteed to appreciate—but only if they remained sealed in protective plastic. After the mark offered up his or her credit card information, a packet of stones would arrive in the mail, but there was no way to verify the authenticity or quality of the product without ripping open the envelope and breaking the rules. The gemstones could be anything from D flawless diamonds to mere junk—and while many people fell for the con, it also didn't always take the mind of a seasoned reporter like Edward Jay Epstein to guess what kind of stones they were.

In the end, the investment frenzy subsided as consumers got savvier and realized that not every stone carried the stunning moneymaking promise of the Taylor-Burton. Exceptional diamonds were genuinely rare, and the global marketplace reacted accordingly. A quarter-carat engagement stone of decent quality would certainly impress a bride-to-be, but as for the resale value—suffice it to say, this didn't rank among the securest investments. But was that ever the point? As time wore on, diamonds returned to what many regarded as their rightful position: something expensive, romantic, shiny, and glamorous.

◇◇◇◇◇◇◇◇◇◇◇◇◇◇◇◇◇◇◇◇◇◇◇◇◇◇◇◇◇◇◇◇◇◇◇◇◇

It helped that there was another royal wedding in the pipeline. After months of "auditioning" for the part of princess by dating Prince Charles in full view of the press, nineteen-year-old Lady Diana Spencer accepted his proposal of marriage. Almost immediately, the media whipped itself into a frenzy describing the diffident young woman who, so far, had behaved flawlessly. She had also maintained an enigmatic distance, which made her the perfect canvas for reporters to color at will. Spencer was beautiful, well-mannered, and well-bred, but by far the most compelling thing about her was that she was the woman who inspired the thirty-two-year-old bachelor prince to settle down. The couple confirmed their arrangement with a most spectacular ring. Garrard, still the English crown jeweler, designed it, and reportedly, the bride-to-be plucked it straight from the existing collection. It featured a large faceted oval sapphire (estimated to be between nine and twelve carats) surrounded by a halo of fourteen diamonds. As the *Los Angeles Times* cheekily put it in the days following the engagement:

"This week Lady Diana Spencer renounced for life her privacy, independence and freedom in exchange for a diamond and sapphire engagement ring and the prospect of a crown."

That the ring wasn't custom-made raised eyebrows only among the toniest English citizenry. The sapphire center stone was huge, and that was enough to satisfy interested parties who knew that a royal wedding was all it took to inspire a fine jewelry renaissance. In fact, it was so big that almost overnight, a rumor surfaced that Lady Diana's future royal in-laws believed she had been immodest to choose it. Just weeks after the engagement became official, a Mr. Stephen Barry appeared on the American television news program *20/20* sharing gossip from Buckingham Palace. He had been Prince Charles's valet for twelve years and then retired a year before the prince proposed. Although it's unlikely he witnessed the event himself—he had already moved on from his royal employ—Barry told the press that Diana had shocked the Queen Mum when she picked out her jewel. In his telling, Prince Charles—who was hopeless at shopping, apparently—arranged with Garrard to offer his young fiancée an assortment of rings to try on. Barry claimed that she paused before pointing to the largest one.

This may or may not have been true. But at a moment when De Beers was realizing that its push to sell melee might have unintentionally backfired by making big diamonds look gaudy, the moment was ripe for a prominent bauble. Diana's pick was lucky for De Beers, even if the featured stone wasn't a diamond. And with a royal wedding on the horizon—the rare event that had even the least fortunate onlookers cheering on the extravagances of the rich—this had all the makings of another landmark era for diamonds. To that end, De Beers announced within industry circles that it would be increasing its advertising budget to $20 million in 1981 ($54 million today), up from $6 million ($16 million) in the previous year. While the campaign would continue to emphasize pieces that required smaller stones—the eternity ring, recently renamed the "diamond anniversary ring," still effectively ate up much of the Soviet melee—it also came freighted with a very specific goal. The diamond industry lived in a house built on the diamond engagement ring, and the foundation could not be left to crumble. Once again, its primacy needed reinforcement. But while they were at it, the aim was to emphasize the magic of larger stones specifically. After all, the ring was a "once-in-a-lifetime" purchase, and maybe young men and women needed a reminder that they had only one shot to get it right.

Thus, in the new ads, "attention will be focused on the diamond and emphasis will be placed on giving the couple a realistic price reference, and a stronger desire for a larger diamond," the *National Jeweler* reported, in an article that ran on March 16, 1981, just two weeks after Charles and Diana's engagement. In a survey, women "ranked diamond size as the *fifth* most important aspect of a diamond engagement ring—on a scale of 1–5—*below* diamond quality, style of setting, shape and color of the diamond." To De Beers and N. W. Ayer, this just wouldn't do. They ran a fresh set of print ads in *Vogue*, targeting women. A beautiful model sat against the backdrop of a teal-blue ocean, wearing nothing, it appeared, except a sparkling necklace. "She can't flaunt a fur on the Côte d'Azur," the copy read, followed by the smaller text: "Born out of fire and ice more than a hundred years ago. Every diamond is unique. But a diamond of a carat or more is even more precious."

"A Carat or More" became De Beers's new refrain, echoed by the heritage tagline "A Diamond Is Forever." In the '80s, the revitalized message was: the bigger the stone, the more depthless the romance. In order to make sure local jewelers maintained a united front at the point of sale, the Diamond Promotion Service sold mail-order training kits with audio and visual elements. For $300 ($810 today), jewelers received a set of tapes, which they could then pass around to everyone on their team, all but guaranteed to increase profits. As long as that happened, the jewelers were happy. But De Beers was happy, too, and not only because it meant that more people were buying more expensive diamonds. These training programs provided another method for controlling the way information flowed to the public. When, in the early '80s, N. W. Ayer came up with the now fabled "two months' salary guideline," the Diamond Promotion Service sold it—hard—and not just directly to the consumers but to the people who would be selling rings as well. One twenty-minute video, which opened with a recording of Harry Oppenheimer's patrician voice discussing the power of a diamond, educated the viewer about the stone's history and science, before introducing the new party line:

"Some of your customers may have little or no idea of what they can expect to spend on this special diamond. Whether these customers are under the age of twenty or over thirty-five, just starting out or already established, the two months' salary guideline will provide a general framework for deciding on a price range, and serve to increase the amount a couple expects to spend on a diamond engagement ring! Now, just as they probably know how much they'll spend on other purchases such as a car or a stereo, they'll know how much to spend on a diamond engagement ring. By making the two months' salary guideline part of your diamond engagement ring vocabulary and your diamond-selling strategy, you can trade up couples to a better quality diamond. In fact, your chances of trading up are better today than ever before!"

The people at N. W. Ayer & Son always did their homework, and they knew that the couples who were getting married concurrently with Charles and Diana were older (an average of age twenty-three for women and twenty-five for men), more successful, and generally savvier. The idea behind the two months' salary guideline was to take advantage of the resources of an upwardly mobile generation and to do so from a posture of goodwill, by giving structure to the conversation around money—always thorny—and also by reminding the pair that "the diamond is a symbol of their love—they'll want a symbol they can be proud of for the rest of their lives." The logic was irresistible, and touched on all the sensitive parts: love, generosity, ego. It gave women license to covet bigger, better stones, and gave men the competitive incentive to buy them.

Out with the hippies, in with the yuppies. For De Beers, the days when lovers opted for pearls, or jade, or nothing at all, were fading into the past.

<hr />

As the economy bounced back in the early 1980s and De Beers worked its magic, diamonds became fashionable again. And not just any diamonds—big, borderline gaudy ones that didn't simply complement an outfit or tastefully enhance a woman's looks, but rather demanded attention in and of themselves. For this reason, it was a good moment for the synthetics market, too. And now, there was a new material available that was so convincing it made Wellingtons look like rhinestones. Supposedly, it even managed to trick the lifelong industry folk at the Diamond Dealers Club, making this product more desirable—and also more disconcerting—than anything that came before it.

It was cubic zirconia, or "CZ."

The Soviets who first synthesized it weren't trying to make jewelry; they were experimenting with materials to use in opto-electronics and lasers. But like many accidental discoveries, this one had unexpected commercial promise. After all, General Electric had only just figured out how to grow gem diamonds, but slow turnaround meant that De Beers was right—as a salable product, it didn't have legs. In terms of simulants, Helen Ver Standig's strontium titanate looked pretty good, but a real connoisseur wouldn't mistake it because the rainbow light it emitted was too bold and flashy for a real diamond. It was soft, too; as charming as Madame Wellington's ads were when they pointed out the absurdity of using one's jewels to cut glass, the truth was that the jewelry made from her simulant wasn't "forever," but in fact was fairly vulnerable to showing signs of wear and tear.

But cubic zirconia was different: It was hard, white, and brilliant, with just the right amount of scintillation. It cut like a real diamond, and that made it easy to fashion well. It was cheap, too. Manufacturers sprang up in New York and New Jersey, as well as overseas in Switzerland and Taiwan. Competition kept prices low, and by 1979, they hovered around $90 a carat ($310 now), compared with the $2,000 or more asked for a same-size middling-quality diamond ($6,884). Skeptics argued that CZ was a fad, with overzealous production poised to implode a promising launch.

However, a few factors suggested otherwise. For one thing, despite the improved economy, the prospect of armed robbery still haunted the rich; as a result, many jewelry lovers preferred to keep their finest pieces locked away in the family vault. Moreover, insurance companies actually raised premiums on diamond jewels if they spent more than a certain number of days per year out of the safe. Owners of exquisite baubles found themselves in a bind: they were financially discouraged from wearing their jewels, and afraid of attracting too much unsavory attention anyway. But what was the point of having jewelry if they couldn't enjoy it? Some people found a solution to this problem with cubic zirconia. They had their diamond pieces copied for a fraction of the price and then wore the CZ versions out to parties, all the while knowing that the real valuables were tucked away in the bank. Even Marylou Whitney (of *the* Whitneys) admitted to doing it—on the front page of the *Wall Street Journal*. Many jewelry stores took the same cautious approach, using synthetic pieces instead of real diamonds in their windows.

There were other, more obvious reasons that customers were interested in diamond simulants. While there was a well-established stigma associated with trying to pass off fake jewels as real, it didn't stop people from doing just that. There were always going to be customers who wanted to look like a million bucks while secretly paying bargain-basement prices. That was the stereotype of a CZ buyer, but it also wasn't the entirety of the market. In the aftermath of the investment diamond boom, some buyers had lost confidence in real gemstones but appreciated the look of a glittering accessory. They were willing to try something new, and unlike the social climbers, they weren't afraid to talk about it. These were the clients that manufacturers of cubic zirconia were most actively courting. Unlike De Beers, Ceres Corporation, a US manufacturer, had no interest in building a mystique around its product. "We make the material as though it were fertilizer," the president candidly told the *Los Angeles Times* in 1984. He went on: "Everybody doesn't drive a Rolls-Royce; a lot of people drive a Chevrolet. I would put CZ in the category of a Chevrolet."

It was a refreshing tactic, and by the mid-'80s, cubic zirconia had become so popular that it was sold everywhere from Sears to Saks Fifth Avenue. But as diamond simulants reached a new level of acceptance in the 1980s, no one would argue that when it came to one particular piece of jewelry, diamonds still had the full advantage. Cubic zirconia and diamonds were like identical twins: bewitchingly similar in appearance and yet distinguished by their fundamental differences.

"We place the emphasis on fun," the president of Windsor Jewels—another domestic distributor of CZ—told the *Los Angeles Times*. "We do not promote our product as a replacement to the traditional emotional attachments people have for diamonds to mark special events—engagements and weddings." And so the engagement ring remained sacred territory, and anyone who thought otherwise was actively shamed. As early as 1957, when General Electric first introduced its industrial diamonds, De Beers and N. W. Ayer & Son got ahead of the message, and headlines like "Man-Made Diamonds Won't Win a Girl's Heart" appeared in the *Hartford Courant*. However, by the time that cubic zirconia was trendy, this idea was so widely internalized that even an unbiased party like Abigail Van Buren of "Dear Abby" revealed strong feelings on the subject. On August 30, 1983, "Debating" confessed that his girlfriend wanted a "big rock," but he couldn't afford it: "Up until recently I was willing to buy her the 'rock' although I am far from rich. Then I saw a synthetic diamond that looked so much like the real thing, most jewelers couldn't tell the difference . . . [which] I can get for about $300."

"Debating" couldn't see the harm in giving his girlfriend what she wanted while staying on top of his budget. But "Dear Abby" stayed firm: "You can't blame a woman for wanting the real thing. However, don't go in hock for a rock. Spend whatever you can—but let it be genuine."

13

The Masters

Australia, Belgium, and India

By the mid-1980s, the diamond trend was robust enough again to warrant a new strategy: De Beers wanted to court more male customers. By that time, it was clear that men were open to wearing some jewelry, from the traditional cuff links, watches, and tie clips to less classic pieces, like gold chain-link bracelets. A few jewelers even tried pushing for a male engagement ring, but it didn't catch on as a style, let alone a new custom. Regardless, the bottom line was that people were making more money again and they wanted to show it off. De Beers and N. W. Ayer sensed a rare opportunity to sell men diamonds.

There was, at least, some historical precedent. Back in the Gilded Age, a self-made railroad-supplies salesman named James

Buchanan Brady stood out among his peer group for his extravagant taste in jewelry. Unlike the other men of his ilk who might choose to wear a lone pinkie ring or a simple jeweled tie pin, Brady showed no restraint, buying up everything—diamond watches, watch chains, cuff links, and pins—and wearing many pieces at once. In this way, he rivaled the American women of his time who layered their finest pieces. He was a brash, conspicuous character within New York's close-knit social scene, and in the society pages was known by his nickname "Diamond Jim."

Brady grew up poor in Lower Manhattan, just a short walk from Maiden Lane, where the diamond district was located throughout his childhood. For his first job, he worked as a messenger boy for the New York Central Railroad. It took years, but eventually he grew up to be vice president of the Standard Steel Company and one of the most famous railroad men of his time. He adored the city's nightlife. Though he wasn't a drinker, his appetite was legendary, and he preferred to indulge in elaborate multicourse meals and then dance into the wee hours, despite his significant heft. Diamond Jim was a lifelong bachelor, and that might help to account for his unusual fashion sense. At a moment when Americans were eager to show off their wealth, the husbands tended to be more subdued while their wives were encouraged to go all out. There was a time when Indian maharajas got their first pick of octahedral Golconda diamonds, and when Louis XIV retained an explorer like Tavernier for the sole purpose of expanding his collection of jewels, but by Brady's day, the post-Enlightenment attitude toward dress had turned most powerful men away from anything that might be interpreted as frivolous and therefore feminine. As the years passed, the men most likely to wear diamonds tended to have a good reason—they worked in the industry. When, in 1921, the *Jewelers' Circular* profiled

William Craig, an engineer for the Kimberley mine, the article noted that even within his circle in South Africa, his penchant for wearing diamonds was unique: "It is a known and recognized fact that many wealthy jewelers and diamond dealers make little display of jewelry for their personal adornment, but not so of 'Sparkling Billy,' whose reputation as a male wearer of precious stones super-sedes that of the late 'Diamond Jim' Brady." Sparkling Billy turned heads everywhere he went, decked out in multiple rings, diamond-studded shirts, cascading scarf pins, and a jewel-encrusted cane.

For men like Brady and Craig, diamonds were associated with a certain level of prestige and accomplishment. But generally, in the twentieth century, the men who wore diamonds for reasons of fashion were considered outliers and there was no sense that they'd eventually become typical. That's why when Gerold Lauck met with Harry Oppenheimer to strategize about how to sell more gemstones, neither suggested convincing men to buy diamonds for themselves. Not until the 1940s, when sales in the United States were so strong that it looked like another Gilded Age, did De Beers make a first, tentative overture toward the men's market. Then, it was to debut a new product: brown diamonds. White diamonds were becoming established as the most desirable gemstone but also, thanks to De Beers and N. W. Ayer & Son, as indelible symbols of love and romance. On the other hand, brown diamonds carried no obvious connotations. Perhaps it was time to create some. For years, the ideal white diamond had been described colloquially by people in the know as being the color of a gin and tonic. Maybe now the amber color was a cue, and brown gems could be compared to something virile and swank, like Scotch or bourbon.

To accomplish this, De Beers drafted a plan that was similar to the one it had successfully implemented for the engagement ring just a few years prior. The cornerstone was a series of print ads. In late 1941, the syndicate announced that a new fall and winter campaign would run once or twice a month in *The New Yorker*, where it would be visible to men of higher income brackets. The ads showed pictures of rings, cuff links, and cigarette cases, all speckled with brown diamonds. The copy, written by the group at N. W. Ayer, was similar in tone to that targeting female audiences in magazines like *Life*, *Ladies' Home Journal*, and *Time*, but it was considerably less flowery: "New York men are wearing diamonds again . . . new brown diamonds . . . attractive gems ranging all the way from the lightness of champagne to the richness of cognac . . . The brown diamond ring is bringing a new note of individuality to the modern gentleman." To reinforce these claims, De Beers had partnered with a hatter called Knox and a prominent New York clothing store, both of which had agreed to do fall window displays showing merchandise in shades of "diamond brown" paired with complementary jewels. Top houses like Cartier, Paul Flato, and Udall & Ballou had all signed on to design brown diamond pieces to be featured in the ads, though they weren't so persuaded by the new product that they were ready to incorporate it into their existing lines. "You can consult your jeweler to have them custom-made," the copy read, implying that no one was ready to take a real chance on brown diamonds until there was proven demand.

The houses were prudent. Ayer and De Beers were a dream team who could sell the proverbial ice cubes to Eskimos, but their 1941 brown diamond campaign was a flop. After it finished out, both ideas—selling liquor-toned stones and creating a reliable male customer base—were dropped. But decades later, as the syndicate made its oversupply of Soviet melee an industry-wide problem, the second question, of whether or not men's jewelry had any traction, arose again. As the PR director of Zales reported back to Ayer in 1962, the Zales shops were having decent luck moving glittering tie clips.

As always, Ayer conducted market research and found that 75 percent of men's jewelry purchases were actually made by women as gifts, and of the remaining 25 percent the vast majority were made by single—as opposed to married—men. The mission then was to bump up the number of men making "self-purchases" of jewelry, and for this, De Beers and Ayer reached out to sports stars from the worlds of golf and tennis to endorse diamonds. The hope was that these refined, manly characters would, by association, make diamonds look both manly and refined. The ads ran in upscale magazines like *Esquire*, *Fortune*, *Town & Country*, and *Forbes*. Pro golfer Ben Crenshaw did one: "I've worked hard to get where I am. And I like to wear things that remind me of my achievements. . . . My diamond makes me feel like I have an extra edge even when I'm not on the course. It's tasteful, understated and extremely masculine."

It was on the nose, and with such an unsteady target, the crystal clear messaging made sense. Interestingly, white diamonds—the very same stones that women wore and coveted—proved to be more appealing to male consumers than brown ones. Apparently, men didn't need their own color scale; glittering white gems would suffice. And why not? While white diamonds could be dismissed as feminine, they also had a long, time-honored record of indicating status. Brown stones, no matter how rich the shade, had generally taken an alternative path: rejected as "off-color" and unloaded into the industrial market, where they were bought in bulk for use in tools, factories, and weapons.

Since that one attempt in the 1940s, De Beers had ceded brown diamonds to their place and didn't bother marketing them. But that didn't mean it couldn't be done.

The remote diamondiferous region of North West Australia had at least one thing in common with the mining area of South Africa: its name. Kimberley, Australia—a wilderness territory characterized by steep mountain ranges and low gorges cut through with rivers—was also named for Lord Kimberley, the first secretary of state for the British colonies. But while Kimberley, South Africa, was the focus of international attention in 1871 when the English changed its name from Vooruitzigt, the Australian Kimberley remained largely ignored until almost exactly a century later. In 1972, a group of five mining companies contributed 20,000 Australian dollars each (about $148,000 in US currency today) to start prospecting for diamonds after a few isolated discoveries signaled there might be a hidden source.

There was no existing infrastructure, so geologists had to be flown in by helicopter to study the soil and riverbeds, looking for trace minerals like garnet, known to indicate the presence of diamond deposits. The work was slow going, especially with Kimberley's monsoon climate, which had only two distinct seasons: wet and dry. Yet findings were encouraging, and in 1976, another mining corporation, CRA (Conzinc Riotinto of Australia), got involved in the company that had come to be known as the Ashton Joint Venture. Three years later, they had homed in on Argyle, in east Kimberley, more than thirteen hundred miles from the city of Perth. Argyle's major pipe is called AK1, and between the fall of 1979, when it was discovered, and 1983, when construction of the mine began, the two companies assessed whether or not the project was economically worthwhile. Not only was building an open pit mine expensive—$450 million, when all was said and done—but it also meant investing in the region so that workers would have a place to live for the duration of their contracts. In this case, there were also other issues to consider, like a vocal Aboriginal population with powerful ties to the land and the previously unspoiled local ecosystem, which would have to be preserved and restored. Ultimately, the numbers still made sense, and by early 1982, news broke across the diamond world that another source, untapped and expected to be extremely prolific, was coming.

Of course, the Australian government and the Ashton Joint Venture had been in communication with De Beers. When construction started, alluvial mining was already under way, as were negotiations, with the outcome considered a foregone conclusion despite Prime Minister Malcolm Fraser's vocal condemnation of apartheid. The Australians expressed real reservations about doing business with any corporation that supported—even indirectly, through tax dollars—South Africa's racist regime. But as everyone in the diamond game accepted, the only way to keep prices stable was to buy into the system single-handedly designed and operated by De Beers. In 1983, the Ashton Joint Venture signed a firm, if unconventional, contract with the Oppenheimers: the CSO would buy and market 75 percent of Argyle's yield, including almost all of the solidly gem-quality diamonds, for five years from the start of operations. The other 25 percent of stones would be Australia's to sell.

This wasn't exactly a coup for De Beers, and not just because it had conceded a quarter of Argyle's output. At the time, diamond sales were still flagging on the heels of two recessions and the damaging—if not altogether devastating, as Edward Jay Epstein had warned—investment diamond craze. The syndicate had already upped its promotional budget, but the industry wasn't nearly on steady ground yet. To protect the value of its most precious resource, De Beers had been furiously stockpiling. The last thing the market needed was a huge mine that spit out unprecedented quantities of diamonds, as many experts were predicting it would. But the now seventy-three-year-old Harry Oppenheimer put on a brave face for the media: "I'm perhaps the only person in our organization who can remember times more difficult than our present ones," he told the *Wall Street Journal* in 1982, probably referring back to his childhood, when his father, Ernest, weathered the Great Depression. "If you have a recession that goes on for five years, you have to do very disagreeable things like closing mines and passing dividends . . . but we aren't going bust, and we're going to hold the market."

He was right. By the time Argyle was commissioned in December 1985, the conversation was different. The global economy had bounced back. N. W. Ayer had convinced Americans that two months' salary was the appropriate amount to spend on an engagement ring, and J. Walter Thompson was pushing for a whopping three months in Japan. However, it was also true that Argyle's yield was daunting. In 1986, Argyle was expected to produce twenty-five million carats—or 50 percent of the world's supply of rough diamonds. Who could say what would happen when the number of stones being mined increased by half? A spokesperson for De Beers, in conversation with *National Jeweler*, kept his tone measured, admitting that yes, while the Australian output was impressive, it was important to keep in mind that the majority of diamonds coming out of the ground in Kimberley were industrial-grade, which the international market could easily absorb. After all, the two biggest producers of synthetic diamonds, used in industrial applications, were General Electric and De Beers, so the syndicate could potentially recalibrate its contribution to make space for Argyle's brown gems.

And a great number were unmistakably brown, anywhere from 50 to 70 percent. The bulk of gems plucked from the local soil were tiny—under two points, or 0.02 carat, after fashioning—and distinctly off-color, ranging from light amber to mahogany. Beyond just being generally concerned about the global diamond market, the people at De Beers weren't terribly enthusiastic about the product, as they made abundantly clear. In truth, this wasn't anyone's idea of a perfect yield, but after all the hard work and money it had put in, the Ashton Joint Venture wasn't willing to simply concede that its stones had no future in the jewelry market. If Ashton eventually wanted to become independent of De Beers—as it most certainly did—it had to find a way to sell these stones for more money than industrials generally brought in. "It was a very fractious relationship [with De Beers]," admits Robyn Ellison, the current communications manager for Rio Tinto Diamonds, who was originally recruited to manage the contractual agreement with the syndicate. "You're selling the majority of your diamonds to your biggest customer and your biggest competitor. So we were in bed with the enemy."

Very quickly, the people behind Argyle realized that in order to make the most of their remaining 25 percent yield, they would have to forge their own way. They opened a marketing office in Antwerp, Belgium, which was still the world's diamond-trading capital, and began sending diamonds westward. However, Belgian customs wasn't entirely prepared for the sheer quantity of stones coming over from Australia. Under normal circumstances, gemstones were shipped in small, easily managed parcels, but Argyle's diamonds arrived by the truckload. The Australian deliveries effectively shut down the customs office as the Belgians processed the stones and tried to figure out how to sort and store them. It was decided that because there was no other place to put the stones, they had to be transported immediately to the new Argyle facility, where a large safe had been installed. The Australian manager offered his own solution: he arrived at customs with a trolley cart and then proceeded to wheel millions of dollars of diamonds through the streets of Antwerp.

The Antwerp office would help with distribution, but there was another pressing issue: whether or not brown gem diamonds would sell. The Ashton Joint Venture wanted to break into the mature US market, and for guidance, it hired a Boston market research and consulting firm that started the project by having samples made up of brown diamond jewelry for women. The pieces all highlighted amber stones, but also emphasized the beauty of contrasting colors: ambers set against white diamonds and then elsewhere alongside deep, warmer browns. The consultants launched an extensive research project that gauged consumer response to this kind of jewelry. When they came back to the client and presented their findings, they were confident. In no uncertain terms: the American shopper would buy it.

But the next step was even more critical: convincing retailers to carry it. For that, the Ashton Joint Venture turned to MVI Marketing, a Southern California firm specializing in jewelry and headed by a husband-and-wife team who both had experience in the industry. Liz Chatelain and Marty Hurwitz first met the clients in 1989, when they flew in from Australia. They had the samples with them and a name already in mind: champagne diamonds. Chatelain remembers first hearing the concept: "My partner and I looked at each other and said, wow, this is going to be difficult.

Only because, one, the trade didn't even know that diamonds came in different colors—they knew the Hope Diamond was blue—but not at this magnitude of quantity were colored diamonds ever exposed to the market . . . and the second issue was that, historically, browns were industrial, and we had that issue to get past with traditional jewelry store owners and retailers." What she meant by "the trade" wasn't fine jewelry houses like Tiffany & Co.; the high-end Fifth Avenue jewelers all embraced colored diamonds. On the other hand, small-scale, local retailers didn't have access to these kinds of rare gems, and if they knew about browns at all, it was only because they had been trained to disregard them as off-color. "We had a joke for twenty years that we don't use the B-word," Chatelain says, "because that was a significant barrier." MVI set up the Champagne Diamond Registry, a trade association funded by Argyle, and introduced a champagne color chart, which provided categories of stones from light champagne to medium to cognac. Then MVI rallied GIA to endorse it, thus standardizing the vocabulary describing brown diamonds and giving everyone in the business a fresh way to talk about them. Finally, MVI enlisted prestigious designers like David Yurman to create pieces using Argyle stones that would be used as samples to entice jewelers.

Over the next two years, MVI's people toured the country, meeting retailers and convincing them to host champagne diamond events at their stores. While some jewelers remained skeptical even after seeing MVI's traveling collection, the customers—in a way that was consistent with the research—proved to be much more open-minded. Without any background in gemology, the average shopper exhibited no bias. MVI supplemented these in-person appearances with print ads in upscale publications like *Town & Country* and *Vanity Fair*. The ads showed real women from various

ethnic backgrounds who looked like professionals—the kinds of
women who might just walk into a store and buy themselves diamond jewelry instead of waiting for men to buy it for them. If this
all sounds a bit familiar, it's because MVI's objective with brown
diamonds was quite similar to the mission De Beers had with white
diamonds in the late 1930s and early 1940s. Just as with N. W. Ayer,
the concept was generic diamond marketing. MVI was promoting,
not Argyle or any particular designer or brand, but rather the idea
of the champagne palette.

Soon, it started working. In 1991, champagne diamonds were
about 30 percent cheaper than white ones of equivalent size and
quality, which meant that champagne was a good fit for playful, trendy pieces that didn't have to be justified as an heirloom
or investment purchase. Because the stones were small, jewelers
started fitting them in multicolored pavé settings that brought out
the contrast between lighter and darker browns. Thanks to Argyle's incredible production, the industry was once again changing
and making room for another category beyond just fine jewelry and
bridal. Fashion jewelry—less expensive items that a woman might
buy just because, or a man might buy her for a Christmas or birthday present—made use of these tiny, lower-value stones.

Also in 1991, the Ashton Joint Venture renewed its contract with
De Beers for another five years, but with an eye toward becoming independent in 1996. Since the Argyle mine opened in 1986, the Australians had demonstrated that they could effectively market their
own stones, even when faced with a supposedly subpar product. "It
was managing your own destiny, the hand we were dealt, which was
this bizarre production," says Robyn Ellison. More, they had started
building their own global relationships outside the syndicate. It
would make them increasingly autonomous as time went on.

Argyle's production wasn't considered ideal, but it came with a silver lining—or, in this case, a mesmerizing fuchsia one. As it turned out, Australia's earth had more to offer than just an overwhelming number of low-value brown stones; it was the first mine in history to consistently produce gorgeous pink diamonds, in small but steady batches. Since the earliest days of diamond mining, pink stones had been extremely rare and were the kind of discoveries that always prompted celebration. No one region was known for producing them; rather, they appeared spontaneously in places like India, Brazil, and South Africa in the same way that blue diamonds or huge diamonds did. Two of the most impressive early specimens—the Darya-i Nur, meaning "Sea of Light," and the Nur ul-Ain, or "Light of the Eye"—are part of the Iranian crown jewels, and are as famous in the Middle East as the Hope Diamond is in the West. Also like the Hope, both of these sizable, softly pink stones have been the subject of considerable debate in the gemstone world and even forensic inquiry into their origins. Gemologists have traced them back to Golconda in the seventeenth century and, more specifically, to the same mother stone: the Great Table, a fabled diamond described by Jean-Baptiste Tavernier in *Les Six Voyages*.

Other pink gems have equally impressive pedigrees and prove all the more enchanting because their histories have been recorded in spurts or, as in the case of the Nur ul-Ain, have gone almost entirely undocumented. In 1990, the Agra Diamond, cushion-cut and weighing in at 32.24 carats, surpassed all expectations when it sold at a Christie's London auction for a record-breaking price of $6,959,780 (about $12.8 million today). The surprisingly high value probably had something to do with its dramatic story, which began in the sixteenth century in India, where it was the treasure of kings. It eventually traveled to Europe—by way of an unlucky horse's stomach, according to one account—and survived Nazi plunder when its owner had it buried, along with the rest of her collection, in her garden during World War II. The legend of the Agra only served to enhance its attractiveness, though a stone of its size and color—a light, slightly orange-tinged blush—would be coveted even without the additional benefit of a good yarn.

But for the vast majority of consumers, the closest they'd ever get to a pink diamond was by buying an Elizabeth Arden nail polish or lipstick of the same name: a promotion originally hatched by Arden and Hollywood jeweler Paul Flato in 1940, after he got his hands on a very large pink that he set in a ring. By 1990, when the Christie's Agra auction took place, pink diamonds were still much rarer than white ones; even in Australia, these stones were scarce, making up one-tenth of 1 percent of Argyle's annual production. Argyle's pinks were truly remarkable, with a rich and saturated color, an almost purply shade reminiscent of an orchid petal—"a fabulous fluke of nature," according to Robyn Ellison at Rio Tinto. Even during the first round of negotiations with De Beers, when the Ashton Joint Venture needed the support of the Central Selling Organization in order to ensure the mine's economic viability, Ashton fought to retain at least a tiny portion of the pink yield that it rightly suspected would become Argyle's signature. Seven years later, in 1991, when the Australians renewed their contract with De Beers, they negotiated to retain all of the pinks.

Rio Tinto sells its prize pinks at exclusive annual events called tenders, the Australian equivalent of De Beers's sights. The inaugural tender was in 1984, when Argyle was still run by the Ashton Joint Venture and collaborating with the CSO. Tenders are global, invitation-only rituals with 150 to 200 participants. They are among the top jewelers, wholesalers, and collectors in the world—in other words, the kinds of people who are expected to buy gemstones that can sell for millions of dollars per carat. Typically, only about fifty carats are on offer per tender, with the individual pink jewels weighing in at about one carat each. These events, scheduled in major cities like New York, Sydney, Hong Kong, and Tokyo, are as much a method of distribution as they are a marketing concept. A rare pink feels all the more exclusive when only the richest, most high-profile connoisseurs can actually get their hands on it.

Ownership of the pinks gave the Australians more clout, especially after celebrities started wearing them. In 2002, Jennifer Lopez famously received a Harry Winston engagement ring from Ben Affleck, with a stunning 6.1-carat light pink square stone and two white baguettes flanking it, worth an estimated $1.2 million ($1.58 million today). But even before that, in 1996, Argyle decided not to renew its contract with the syndicate. What this meant was that all of those small brown stones would be the Ashton Joint Venture's exclusively to market, and luckily, the champagne diamond trend had gathered a fair amount of momentum. More important, the Australians had invested in the Indian diamond-cutting center. Now, when thirty million carats—or almost forty-three million, at its peak production—came out of AK1, there was somewhere to send them for inexpensive processing. In India, diamonds could be fashioned at a small fraction of the price commanded in places like Belgium, New York, or Israel.

India had an ancient cutting tradition dating back to the days of Tavernier, but the art form hadn't evolved as it did in America and Europe. When the Australians started funneling their diamonds to India, that deficit was apparent: the quality of the work was noticeably inferior to the standard coming out of competing areas. The Ashton Joint Venture decided to train Indian cutters to make the kind of pieces that would sell overseas, which had a dramatic effect on the worldwide industry. India provided a reliably low-cost corridor for small, flawed diamonds; it was a place to make brown diamonds into baubles, but also where a very light champagne stone could be cut and set in such a way that no layperson would be able to recognize it wasn't white. The idea wasn't to trick consumers into purchasing off-color stones, but rather to make diamonds so affordable that anyone who wanted one could have it. It worked: the average price of a tennis bracelet in the United States, in 1984, was $6,000 ($13,800 today). Five years after Argyle unleashed its enormous production into the global marketplace, the average price dropped to $300 ($580). That's because the stones were lower quality and processing in India was significantly cheaper. Suddenly, even American big-box stores like Kmart, Target, and Walmart could afford to carry diamonds. It also meant that the people who shopped in places like Kmart, Target, and Walmart could afford to buy them.

Diamonds the world over are cut either by a single artisan—an honor usually reserved for the most precious gems—or by a series of cutters, or by some combination of human and machine, or even entirely by machine in some cases. Today, it's not necessarily the size but the quality of the rough that tends to dictate where in the world it lands. Diamonds are directed through different pipelines depending on their intrinsic potential. Showstopping stones will always fall into the hands of a master, and every aspect of the process from cleaving to polishing will be handled by an artist. But that's rare, and even more average-size high-quality ones—destined for engagement rings or fine jewelry—will be treated differently from Argyle's typical yield.

That's not to say Argyle wasn't a desirable holding. On the contrary, when the opportunity came up in 2000 for De Beers to make a takeover bid, it offered $388 million—only to lose out to Rio Tinto, the underbidder at $371 million, after De Beers withdrew on account of regulatory delays. For the first time, De Beers had a formidable competitor, and that would require some degree of recalibration.

14

The Stunners

HOW TOUGH GUYS STARTED WEARING DIAMONDS

United States

I n 1979, the same year that the AK1 pipe was discovered, a young man named Yakov Arabov immigrated to New York City with his family. They were Bukharan Jews from Uzbekistan, then a part of the Soviet Union, and they arrived in the States with very little money. Arabov, just fourteen years old, enrolled in high school but felt compelled to help his parents cope with their financial troubles. Back at home he'd apprenticed with a talented photographer who noted Arabov's comfort behind the lens. The boy had a knack for spotting a good subject and imagining what it would look like frozen in place. Now in the United States, the young teenager wanted to build a professional career even before he graduated. He considered becoming a photographer or a hairstylist—another job that would make use of his visual sensitivity—but he signed up for a six-month government course in jewelry design. Arabov saw the potential for growth in that industry; he had four sisters, and he noticed the way their faces lit up around pretty little trinkets. He also felt opinionated about what they wore, and if given the opportunity, he knew exactly what he would change.

After learning how to use the tools of the trade, Yakov Arabov anglicized his name to Jacob Arabo, left school, and found a job at a jewelry factory where he earned $125 a week (approximately $338 today). It was a good salary for a sixteen-year-old boy, but Jacob wasn't there for just lunch money; he had grand ambitions of earning enough to take care of his entire family. Although he was hired for production, he started experimenting with leftover metal scraps and asked his boss if he could try his hand at design. The factory supplied jewelry to chain stores, and successful designers were measured by the number of orders their work attracted; his boss couldn't see any harm in letting a determined kid fool around with the excess stock. And so Arabo started experimenting, twisting narrow tubes of gold or silver into unusual earrings that caught the attention of buyers. His items weren't branded, but within nine months, everyone at the factory knew Arabo was one of its top-selling designers. His paycheck jumped to $1,500 a week ($3,700 now).

It was a huge amount of money, especially for a person his age, but Arabo understood instinctually that the real entrepreneurs— the ones who didn't have to answer to anyone and who had license to create whatever fantastical pieces they might imagine—worked for themselves. At the age of seventeen, less than a year after he started designing, he shocked his boss by giving notice, aware that the longer he stayed, the harder it would be to walk away from such a good, steady income. In his brief time at the factory Arabo had saved over $10,000, which he invested in his own private manufacturing facility. He didn't have the resources to make bulk jewelry up front, so he started working as a contractor, contacting brands and offering a deal wherein they purchased the raw materials and he designed and manufactured items for them to sell. It worked, and within just a few months he had hired ten employees. He then invested the profits from that business into making his very own line, which he pitched to department store buyers in person, using his free time to drive around New York's tristate area with his wares packed into a suitcase.

Like Harry Winston—whom Arabo counts as an inspiration—he initially had trouble making connections in the industry. He was a dark-skinned high school dropout with a heavy Russian accent, and there were very few people who took him seriously. To counter that, Arabo invited his much older second cousin, also named Jacob, to join him as a partner. Together, they founded J+J Jewelry Company and settled into a routine with a clear-cut division of labor: cousin Jacob handled the clients, and Arabo toiled behind the scenes. A few years later, he made the decision to shut down his manufacturing firm. It wasn't creative, and at twenty years old, he'd already accomplished his major goal: he bought his family a house. He was supporting his parents and his sisters, and with the leftover profits, he felt entitled to pursue another, more abstract goal: true artistic fulfillment.

The two Jacobs opened a small retail stall on Sixth Avenue and Forty-Seventh Street, in the heart of New York's diamond district. Arabo displayed his creations in the window, and they immediately stood out from the others on the block owing to the unusual size and scope of his designs. Unlike many of his neighbors, he didn't come from a jewelry background, and that meant he didn't have the industry contacts or the financial safety net to keep his business afloat in rough waters. His competitors told him he was wasting his time with showpieces; rather, regular income came from walk-ins, usually men in the market for diamond solitaires and young couples shopping for wedding rings. It was rare to sell anything else. But Arabo hadn't come all this way to spend every day dropping white diamonds into knockoff Tiffany Settings, and he wouldn't let a few naysayers scare him. Arabo remembers: "Before I went into the business my friends told me, don't, what are you doing, there's sharks all around you, they're gonna eat you alive. I said, I guess I have to become a shark."

Jacob Arabo had already cut his teeth. The next step was learning to hunt.

◇◇◇◇◇◇◇◇◇◇◇◇◇◇◇◇◇◇◇◇◇◇◇◇◇◇◇◇◇◇◇◇◇◇

Arabo opened his store in 1986. What he didn't yet know—but would soon find out—was that his designs were actually a perfect fit for a certain population of high-profile, up-and-coming jewelry lovers who appreciated the kind of unique, outrageous, and expensive pieces he was displaying on Forty-Seventh Street. As Arabo spent his nights at home, laying diamonds out on white sheets of paper and envisioning how they'd inspire his next creation, a group of people heretofore ignored by the jewelry industry were in the process of rerouting the worldwide conversation about gemstones. They were an entirely unexpected segment of the US population: African American men. By the early 1990s, rappers were becoming music industry powerhouses and unquestionably legitimate cultural influencers who were delivering their own unique flair to the masses. In doing so, they transformed the diamond's profile so that a very costly, high-quality stone was the gem of choice not just for blushing brides-to-be, actresses, and society doyennes, but also for black and Latino men who had reached a certain rung of affluence. Interestingly, the majority of white men at the time weren't wearing diamond accessories anymore—that moment had passed along with the 1980s. But this trend had its own distinct evolution, moving on a track largely parallel to but also entirely independent from mainstream fashion. Hip-hop was a movement that began in the late 1970s and early 1980s with a distinct sound but also its own unique dance moves, vernacular, visual art forms, and sartorial point of view.

"It should never be a surprise that a lot of the status symbols that equate with success in the broader American culture also equate with success in the urban community," media personality Bevy Smith points out when discussing why African American men initially gravitated toward diamonds. Successful musicians in the 1970s wore them—Isaac Hayes, Barry White, the Isley Brothers—and even before that, there was Sammy Davis Jr. in his eye-catching necklaces and rings. In the 1970s, fashions for men were more adventurous and androgynous than ever before, but those album covers, Smith explains, likely stuck out in the memories of the next generation of influential, jewelry-clad black male artists like Slick Rick and the guys from Run-D.M.C. Back then, yellow gold was the precious metal of choice, which was particularly impressive given the outrageous price jump that occurred in that market in the 1970s and 1980s. Slick Rick was known for wearing as much heavy gold jewelry as he could physically carry, like layered chains with links the size of bottle caps and bigger, and pendants depicting everything from religious symbols to—embossed on a flat, round medallion that dangled down to his waist—an image of the scales of justice. Flavor Flav, from Public Enemy, who had a similarly exaggerated look, was known for the elaborate clocks he wore around his neck like a character out of *Alice in Wonderland*. By comparison, the trio from Run-D.M.C. were almost understated in their signature fat gold rope chains and watches, which still looked as though they cost a pretty penny.

The referent for this style wasn't just old-school soul singers: it was kings. In the same way that at the turn of the twentieth century, New York's industrialists and their wives looked across the ocean to European royalty for their inspiration about how to convincingly portray their wealth, hip-hop artists did, too, eager to show off their success. Between the Gilded Age railroad baron and the late-century rapper, there's an unexpected through line: the nagging insecurity that often comes with new money. The difference is that, with the exception of Diamond Jim, the industrialists had relied on their wives and their homes to be reflections of their growing fortunes, keeping their own dress relatively simple. Even if, deep down, the men wanted to wear opulent jewels in the fashion of Louis XIV or Tutankhamen, they were generally restrained by the conventions of high society, which prescribed that men look serious and dignified. On the other hand, hip-hop artists were often single, and even if they weren't, they were the stars, the magnificent peacocks appearing onstage and, later, on MTV. Slick Rick was even known to wear a crown.

DJ, brand manager, and hip-hop entrepreneur Peter Paul Scott credits influences as diverse as the internationally beloved performer Liberace and marginal urban pimps and drug dealers for shaping the blinged-out look of hip-hop in the late 1990s and beyond. Again, the commonality is a disadvantaged background—Liberace was the son of working-class immigrants—followed by hard-won wealth, and that sharp contrast creates the impulse to compensate, to visually distinguish between the selves of past and present. As Scott points out, pimps and drugs dealers are local celebrities, the guys from the neighborhood who can prove that they've made it—specifically, through their access to girls, cars, nice clothes, and jewels. It makes a lasting impression, especially on kids who have nothing. When those same kids grow up and get rich, they have a crystal clear sense from the old days of what that experience should look like.

"With the hip-hop community, we remix the way things are used," Bevy Smith explains. Music producers sampled existing songs and gave them new beats, new meaning, and new life; in the same way, rappers took mainstream status symbols and made them their own. In Harlem, New York, in the 1980s, a haberdasher who called himself Dapper Dan made fabrics covered in high fashion logos like Gucci and Louis Vuitton and turned them into things like one-of-a-kind leather jackets and tracksuits. Gaudiness wasn't something to avoid; quite the contrary, it was the whole point. With the arrival of crack cocaine in the region, a growing number of drug hustlers had money to spend on custom outfits. Dapper Dan had plenty of legitimate clients, too, performers like LL Cool J, Salt-N-Pepa, and Biz Markie—who name-checked Dan in his song "Throw Back"—and heavyweight boxer Mike Tyson.

In the late 1980s, major record labels started investing in hip-hop, and suddenly, what had been an underground movement—with the exception of a handful of breakout acts—turned mainstream. MTV promoted airwave-friendly rappers like MC Hammer and Vanilla Ice, but also took its chances playing controversial videos from the so-called gangsta rappers like Dr. Dre and his protégé Snoop Dogg, the Notorious B.I.G., and ex–N.W.A member Ice Cube. The gamble turned out to be a very good one, and in 1990 and 1991, the music industry saw a sizable increase in profits with a decent percentage attributed to the rising popularity of hip-hop, roughly 9 percent of a total $7.8 billion in domestic sales. This meant that the artists were reaching wider audiences than ever before and that their bank balances were expanding commensurately, with little dynasties erected around the country.

Producers and artists took great pride in their label affiliations, treating them almost like invitation-only clubs, gangs, or sports franchises. The East Coast–versus–West Coast feud that is said to have claimed the lives of Tupac Shakur and Biggie Smalls had the intensity of an ancient rivalry, recalling the Hatfields and the McCoys, the Montagues and the Capulets. Artists and producers often expressed their loyalties by way of their jewelry. Death Row Records CEO Marion Hugh "Suge" Knight Jr. and rapper 2Pac wore matching gold chains with large pendants depicting the label's morbid logo in diamonds: an inmate strapped to the electric chair. Cash Money Records cofounder Bryan "Birdman" Williams commissioned an enormous rectangular pendant with the label's name spelled out around a sparkling dollar sign. But no crew treated a chain quite like the guys from Roc-A-Fella Records, a New York–based subsidiary of Universal founded by Jay Z, Damon Dash, and Kareem "Biggs" Burke. Receiving a diamond-studded Roc-A-Fella chain—a round record with a large R, also representing the official logo—was a rite of passage for recently signed artists, signifying their official arrival into the family.

When Kanye West, a young producer with dreams of being a performer himself, finally signed a deal in 2002 to make his own album, his entrance into the stable of Roc-A-Fella rappers was commemorated publicly by his colleagues. At a concert that August, they gathered around him onstage and presented him with his very first chain, an item that Los Angeles jeweler Ben Baller estimates cost about $28,000. The scene almost had the feeling of a coronation, and it was so significant to West that he included a recording of it as the culminating sequence in his 2003 video for "Through the Wire," his first hit as a solo emcee. For a guy on the Roc-A-Fella roster, the chain wasn't all that different from an engagement ring: it was at once a trophy and an invitation, indicating the end of courtship and the beginning of marriage.

But not all hip-hop chains carried so much symbolic weight; others were what the diamond industry would call "self-purchases," bought to make the owner look rich, or unique, or intimidating. Rappers gravitated toward traditional industry symbols like crosses and "Jesus pieces," but they also one-upped one another with extravagant designs that ranged from intentionally shocking—Shawty Lo wore a "crack vial" that took advantage of the diamond's resemblance to another kind of white rock—to playful and silly, like a PlayStation 3 controller, Bart Simpson, or a box of Crayola crayons. Bevy Smith describes how a young, newly wealthy rapper might end up going home with something like a gem-studded Garfield pendant: "People had the most magnificent cartoon-character symbols. . . . It was like being a kid in a candy store, but the candy wasn't peppermint sticks, it was diamonds. That was the most amazing chance to let your imagination run free, and of course we know rappers have the greatest imaginations, they are some of the best modern-day storytellers around."

If going to the jeweler was like taking a trip to the candy store, then Jacob Arabo became Willy Wonka: the ultimate candy man. One day, after people in the entertainment industry had just begun taking notice of the Russian jeweler's inimitable and unrestrained style, the rapper Biz Markie came to him with an odd request. He wanted a signature piece, something so big and eye-catching that it worked as a costume: in other words, something fans could see from the stage. Arabo appreciated the challenge and got to work, designing a four-finger ring that spelled "Biz" in diamond-studded script. "Nobody back then would make something like that," Jacob recalls. The performer loved it, wearing it not just at concerts but also for photo shoots and on the cover of his hit single "Just a Friend." After that, Arabo's custom business took off. He made his clients comfortable: despite his exquisitely tailored suits and throwback pomade waves, he, like most of the rappers and athletes he catered to, still had the faint shadow of the outsider, a memory of the time when the path to success was still poorly lit. In the end, maybe diamonds made the road a little brighter.

But Arabo was also more flexible than the average jeweler; following in the steps of his idol Harry Winston, he would let his clients walk out with million-dollar baubles before they paid him. He respected his customers and trusted that they'd make good on their balances. And they always did—Arabo also accepted cash payments without asking too many questions and, more important, was unusually lenient when it came to letting clients trade pieces in. After all, rappers were public figures who could get away with wearing their latest purchases only so often before they lost their impact. Arabo accommodated them by taking commissions and then, when the time came, accepting old pieces back in trade as long as the new job was more expensive than the previous one. He never repurposed his designs, appreciating how highly prized the expression of individuality was within his target market of music industry celebrities and sports stars.

In the mid- to late 1990s, at the height of his success, Arabo— who earned himself the unforgettable nickname Jacob the

Jeweler—received countless lyrical acknowledgments, also known as shout-outs, from artists at the top of the game like Jay Z, Kanye West, Fabolous, and R. Kelly. "I even met some of these singers *after* they sang about me," Arabo says. "That was shocking." These mentions did more than just flatter the jeweler's ego; they also boosted his public profile so that other up-and-coming performers knew whom to call when they got their first big payout. Arabo had never invested in marketing, and the songs were the best kind of free advertising. Not only did they attract other entertainers; they brought in average rich folk as well. Kids who heard his name on the radio would encourage their parents to go to the store to meet the famous Jacob. Arabo's Forty-Seventh Street stall became a revolving door of recognizable faces, including his own as he was photographed at exclusive parties and red-carpet events. "That's how big [Jacob] got," Peter Paul Scott remembers. "Not only did you walk around with the jewelry, you walked around with the guy."

∞∞∞∞∞∞∞∞∞∞∞∞∞∞∞∞∞∞∞∞∞∞∞∞∞

Mainstream American viewers grew accustomed to seeing men— and tough, foul-mouthed, tattooed ones at that—wearing the kind of jewels that would make even Elizabeth Taylor's tastes seem minimalist. A gold Rolex started around $10,000 and then the price of customization—adding a ring of diamonds called a bezel around the watch face—made it exponentially pricier. Beyond the typical baubles like watches, stud earrings, bracelets, and rings, rappers frequently wore massive gold or platinum chains around their necks, often featuring elaborate diamond-studded pendants with creative designs. By the late 1990s, this kind of jewelry had become so fashionable in hip-hop circles that Jay Z, who had his first main-

stream hit with "Hard Knock Life (Ghetto Anthem)" in 1998, released a single called "Girl's Best Friend." The track was a cheeky
love letter to diamonds, as spirited as the song performed by Marilyn Monroe so many years ago. Jay Z revealed himself to be a connoisseur: "When you blush you turn blue if your grade is right." In
his telling, the "thugs" loved gemstones just as much as the ladies
did and that addiction was costly, in much the same way that maintaining relationships with multiple women can very quickly take
its toll on a Casanova's bank account.

The practice of flaunting one's opulent lifestyle had become par
for the course in the rap community: the Notorious B.I.G. and Puff
Daddy filmed the video for "Hypnotize" while drinking champagne
on the deck of a yacht, and even Dr. Dre, who made a name for himself as a raw voice straight out of the low-income city of Compton,
had already undergone a rags-to-riches makeover in his 1997 song
"Been There, Done That," which enumerated Dre's considerable
possessions: cars, a private jet, a mansion, diamonds. He'd traded
the Los Angeles street-life references for ones that felt, at least in
hindsight, more epic and glamorous: the Italian Mafia and, more
specifically, Lucky Luciano.

In March 1999, the New Orleans rapper B.G. (Baby Gangsta) released a single with label mates Birdman, Juvenile, and Lil Wayne,
all signed to Cash Money Records. The song had a bouncy, upbeat
groove, and the accompanying music video made use of various
props that, in the preceding years, had become the typical trim-
mings of the genre: girls, cars, helicopters, and jewelry. "Medallion
iced up, Rolex bezelled up, and my pinkie ring is platinum plus,"
Baby boasted. As the guys rapped, they indicated their diamond
earrings, bracelets, and grills. They were wearing a fortune in jewelry, and they wanted the world to know it; it was an effective way of

showing their audience, from the competing talent to pimple-faced white teenagers watching MTV from their suburban homes, how successful they were.

There was nothing particularly novel about a bunch of tough guys rapping about their jewelry. The girl's best friend was already doing double time as the rapper's flossy mistress. What made B.G.'s video memorable wasn't the content or even the song's catchy beat, but the slang it introduced to a whole new audience: "bling-bling."

"Bling" is an ideophone describing the cartoon sound that might happen when light reflects off a diamond. It's the acoustic double of the words "sparkle," "scintillate," and "glitter"; if one could listen to a stone's brilliance, the noise it would emit is *bling-bling*. But the meaning can drift as well. A diamond itself is "bling," as is a diamond-studded jewel, thus the common usage: check out my bling. All of this seems obvious now, but at the time, the slang was eagerly embraced by mainstream America for its almost Shakespearean inventiveness. The expression became shorthand for a trendy, moneyed aspirational lifestyle portrayed by rich African American rappers but also by music industry executives, actors, and sports giants. Superstar Lakers center Shaquille O'Neal was responsible for the expression's first official appearance in the *New York Times* in 2000, when he told a reporter that he had his eyes on the championship trophy: "I want the bling-bling." A few months later, the paper of record officially defined the term for its readership in another basketball profile, this one about freshman Knick Glen Rice, an ex-Lakers player. Reporter Thomas George described his championship ring from his time in Los Angeles: "So many blinding diamonds and the words 'Bling! Bling!' were inscribed on the side, hip talk nowadays for the coolest, shiniest diamonds."

It wasn't the first time that diamonds had received their own

special nickname. In the United States in the 1920s, freewheel-
ing flappers called their diamonds "ice." They also used the term
"rocks" for the big, shiny ones, but "ice" was more popular—a word
that suggested something cool and slick. Jazz Age gangsters and
their molls loved their strange slang, and not only because it was
slippery vocabulary used to slide their true meanings past law en-
forcement officials. That was part of it, but slang was also valued as
a way of distinguishing those in the know—young, illicit partygoers
in the Prohibition era—from everyone else: generally old guard,
religious types. A woman who felt at home at the gin joint knew
exactly what to call her sparkling diamonds. The language of the
speakeasy was just as important in that world as the hairstyles, the
dance moves, and the fashions, differentiating regulars from mere
visitors. It made sense—at a moment when one's choice of hangout
had significant moral and even legal implications, it was smart to
be wary of outsiders.

But this time, the language of the underground was co-opted.
Black rappers took a classic status symbol and then "remixed" it—to
use Bevy Smith's word—and white audiences liked the way the new
version looked and sounded. In 2002, CNN anchors were encouraged to use slang like "bling bling," "fly," and "flava" to appeal to the
ears of younger audiences.

For the first time in decades, diamonds were not only considered glamorous, romantic, and prestigious; they were also down-
right cool. This was great news for the people at De Beers, who in
all of their years of spearheading generic diamond marketing had
never been so bold as to attempt an association between the stones
and the ideas of being tough or cutting-edge or even, really, hip. In
other words, if Jacob Arabo was benefiting from free publicity that
his famous clients offered, so, too, were the diamond sellers.

Meanwhile, Arabo's career continued to grow. In 2001, he debuted his first timepiece: the Five Time Zone watch, an accessory for the international businessman and the true (or wannabe) jetsetter, with four clock faces embedded within the primary one that simultaneously showed the hour in New York, Los Angeles, Paris, and Tokyo. He had dreamed of having his own branded watches, but although breaking into the New York jewelry industry had been difficult, it was nothing compared with the challenge of winning the respect—and the business—of the watchmakers in Geneva. For a long time, Arabo had been content to customize his clients' existing timepieces, adding gemstones or, in some cases, removing the subpar ones the original designers had set in the bezel and replacing them with extremely expensive, high-quality jewels. But then one popular company, called TechnoMarine, served him a cease and desist order. Arabo signed a document agreeing never to touch its products again—and vowed to make his own "Jacob" watch, which quickly became a status symbol to rival the Rolex.

Unfortunately, that wasn't his last brush with the law. In 2008, Arabo was sentenced to thirty months in prison after he pleaded guilty to charges of falsifying records and lying during an ongoing federal drug investigation of Detroit's "Black Mafia Family." After serving his time, Arabo returned to running the business Jacob & Co, though his focus moved away from hip-hop showpieces and toward high jewelry and red-carpet glamour, leaving his old job title—the King of Bling—open. Ben Baller was more than willing to take up the mantle. The owner of the Los Angeles store IF & Co. ("Internally Flawless," though some customers evoke the slang "Icee Fresh") is widely regarded as the "New Jacob."

Baller, born in LA's Koreatown and originally named Ben Yang, comes from an immigrant family like Arabo; his mother arrived in California with just $300 in her pocket. Growing up poor in the 1980s, the young Ben found that gold carried special significance for him. His uncle was a jeweler, and once his mother made some money, she bought herself two gold chains: one herringbone, one box-link. When an exhibition of King Tut—the legendary Egyptian monarch buried in an elaborate gilded sarcophagus—came to town, Yang went to see it twice.

In the mid-1990s, Yang's cousin took over his father's jewelry store while Yang—who earned himself the nickname "Baller" as the only Asian guy playing competitive basketball at his college—started DJ-ing. He was a charismatic kid with a healthy dose of bravado from his days on the court and an appetite for celebrity, and he parlayed his nightlife connections into a job in the music industry. Before the age of thirty, he had worked his way up to vice president of a major record label. He'd also gone broke—twice. Even with a six-figure salary, Baller couldn't make ends meet, spending a fortune on grown-up toys like cars, jewelry, and his biggest weakness at the time: limited-edition sneakers. To pay the bills, he brought one of his lesser gold chains to his cousin to see what he could get for it. His cousin, to Baller's surprise, offered $1,100. Suddenly, the guy with a prestigious record industry title realized that, compared with the jewelry business of his cousin, who drove a Corvette and a BMW, music was low on cash flow. Baller quit his job, sold his rare sneaker collection—for a whopping $1.2 million—and left Los Angeles to travel.

While visiting Tokyo in 2004, he saw a display of hip-hop jewelry and got in touch with his cousin. He realized each of them had something to offer the other: not only did Baller have a fine arts degree and want a job that inspired him, he also had connections with people in the music business who bought and wore high-priced jewelry. His cousin, on the other hand, ran a successful store, but was struggling to attract more prestigious customers. When Baller got back to Los Angeles, the pair partnered up, changed the company's name, and started aggressively targeting entertainers. Baller used the then-thriving social media site MySpace to market himself, and won commissions from the guys in the band Blink-182 and from the duo called Clipse, who were produced by Pharrell.

Baller was determined to make IF & Co. the go-to designer for entertainers and hip-hop artists. He was outgoing, with a shrewd understanding of branding from his time in the worlds of music and nightlife, and he had a sense of what it would take to become a celebrity in his own right, à la Jacob the Jeweler. "I was like a dude out of jail," he remembers. "I wasn't leaving any stones unturned, I was just going after anyone and everyone. . . . At a certain point, I got a hold of Michael Jackson and it started changing." Designing a piece for the King of Pop gave Baller confidence, and most certainly raised his profile, but the real game changer was designing for a teenage YouTube star with great hair, a Jackson Five—era voice, and precocious talent for exploiting social media: Justin Bieber. Bieber and Baller became friends, and the heartthrob put the weight of his publicity machine behind his new favorite jeweler.

After that, everyone in the music industry wanted a bauble from IF & Co., and Baller developed a celebrity-consultation style. If clients are shopping for a signature pendant, Baller starts by teasing out their preferences. Do they want platinum, rose gold, or yellow gold? Do they have a special logo or a family crest? Are they envisioning a necklace, a brooch, or a bracelet? Baller says that, at this point in the process, no inspiration is too outlandish: "Give me all your ideas, I don't care if it's stupid or cool. From there, we start narrowing down." Eventually, he helps each client decide on

an image, and then they work together figuring out how to bring it
to life, drawing from a mix of white diamonds, colored diamonds,
and other precious and semiprecious gems. Once the design is finalized, he sends it to a wax cutter to make a model. Meanwhile,
if the customer has settled on a chain, they need to choose the
length—it can hit anywhere from midchest to the solar plexus to
the belly button—and the link style. When the wax model comes
in, the client gives final approval and then it goes into metal casting. The resulting gold pendant gets polished and set with stones
until it's ready for delivery: "I deliver and people's eyes are like,
oh wow, man, fuck, you made it happen. It's amazing."

Ben Baller gets lyrical shout-outs just as Jacob the Jeweler once did:
his name appears in songs by Wale, Tyga, and Pusha T. For young guys in
the hip-hop world, niche jewelers who have earned this kind of credi-
bility are appealing for a few reasons. Even though rappers are often extremely brand conscious, name-dropping everything from luxury cars
to high fashion houses to top-shelf liquors, they also prefer working
with people who make them feel comfortable. That's where the Jacobs
and the Ben Ballers of the world excel. Baller, who has known both ex-
tremes of poverty and great wealth, admits that even for him—a guy who
makes a habit of walking around in hundreds of thousands of dollars in
jewelry—an establishment jeweler like Cartier can feel unwelcoming.
The best hip-hop jewelers have the talent and access to high-quality
materials, but they also relate to a client who wants something gaudy, or
ridiculous, or both. Probably equally important, they charge less than
the legacy brands. Ben Baller considers himself a "concierge" who will
frequently go to clients' houses if that's where they want to meet. "We
bring Cartier to them," he says of IF & Co., "with our own little hip-hop
touch."

A new, more diverse generation of consumers would grow up

soon, raised on a diet of flashy music videos and shows like MTV's *Cribs*, which was an updated *Lifestyles of the Rich and Famous* for the music world, bringing cameras into chart-topping artists' multimillion-dollar homes—and their jewelry boxes. All of this was, at least in theory, terrific news for the diamond industry. But then there was other news, not out of New York or Los Angeles but from far-off places like Angola and Sierra Leone, that threatened to engulf it.

The Critics

Worldwide

Back in 1998, a year before B.G. released "Bling Bling" out of New Orleans, a relatively young nongovernmental organization (NGO) called Global Witness published a fifteen-page report called *A Rough Trade: The Role of Companies and Governments in the Angolan Conflict*, which would eventually help to do something that very few had managed since the days of Cecil Rhodes: it changed the way the diamond business operated. *A Rough Trade* focused primarily on Angola, a large southwestern African nation that had been mired in violent conflict since its war of independence against Portugal started in 1961. When the colonists pulled out in 1975, they left a vacuum of power, and a bloody struggle erupted almost immediately between two parties: the MPLA (the People's Movement for the Liberation of Angola), backed by the Soviet Union, on one side and UNITA (the National Union for the Total Independence of Angola), backed by the United States and China, on the other. Two wars and hundreds of thousands of casualties later, the vicious UNITA had lost a national election to MPLA but persisted in terrorizing its countrymen, particularly civilians.

Then, in an effort to finance its ongoing rebel campaign, UNITA seized control of an estimated 60 to 70 percent of Angola's considerable alluvial diamond fields, where high-quality stones could be found using primitive, inexpensive methods: digging a few feet under the earth and then washing and sifting to uncover them. This meant that the profits from selling Angola's stolen diamonds went to financing UNITA's continuing terror strike.

In 1994, the United Nations intervened with the Lusaka Protocol and established a joint commission comprising the UN, the remaining shreds of the rightful Angolan government, and even representatives from UNITA, provided the rebels disarmed. Three years later, fighting was still under way and the UN Security Council imposed a first wave of sanctions against UNITA for repeatedly violating the cease-fire. When that didn't stop UNITA, the Security Council unanimously passed Resolutions 1173 and 1176, which expanded the sanctions, this time to include diamonds. The new resolutions prohibited member states—which included the diamond capitals of Belgium, Israel, India, and the United States—from importing any diamonds from Angola that weren't government certified, meaning clearly originating from the remaining sliver of legitimate government-controlled areas. The idea was to cut off the international financial blood flow to UNITA and thereby cripple it.

But the rebel faction proved to be extremely resilient, and Global Witness, a London-based organization that explores the connection between the presence of natural resources in unstable areas and devastating conflict, zeroed in on Angola's diamonds and the way they moved through the world. The NGO initiated its own inquiry into the question of whether Angola's diamonds were being used to finance or even provoke human rights violations. Its findings, published in *A Rough Trade*, were unequivocal: "To the millions of Angolans who have survived the repeated years of conflict and the estimated 300,000 that died violently between late 1992 and 1995, the workings of the international diamond trade may seem an abstraction, but the revenue that UNITA has been able to generate from diamonds and the direct and indirect impacts of this revenue are real enough." It argued that Resolutions 1173 and 1176 did not go far enough, especially given that—and this was the most damning pronouncement—the international diamond community was prepared to skirt any and all limitations imposed by the UN.

The diamond industry, Global Witness reported, had functioned without any real regulation for too long, and this left major players like De Beers in a position to call all the shots. Expecting a multi-limbed capitalist corporation to willingly act against its own economic interests in order to protect the safety of the Angolan people was optimistic, at best. Global Witness conducted interviews and turned to De Beers's own annual reports to show that Angolan diamonds contributed to the syndicate's significant profits and then outlined the various ways the industry had bypassed sanctions.

The first way was through De Beers's outside buying offices. The company had three principal methods of acquiring stones: first, through its own mining operations, such as those in South Africa and Namibia; second, through government contracts to buy and market diamonds, such as the contracts it had with the Soviet Union and Australia at the time; and third, through buying offices stationed around the globe that functioned as trading posts. This last method was by far the most difficult to monitor, as sellers showed up with stones of questionable provenance. After the UN Security Council issued its verdict on Angola, UNITA was still able to smuggle diamonds through a complex network of dealers, making it very difficult to trace their origins. However, Global Witness contended that both these diamonds themselves and their typical chain of possession had certain undeniable trademarks and that the open-market exchanges simply allowed those involved to turn a blind eye to the fact that they were likely handling illegal Angolan gems.

A Rough Trade also singled out the city of Antwerp, still the diamond-trading capital, for careless behavior when it came to spotting illicit imports. A great number of UNITA's diamonds that came through Belgium had done so with incomplete or wholly forged certificates of origin, and Global Witness pointed out that customs agents were consistently neglecting to flag them, even when the country listed shared a border with Angola and had no known naturally occurring mines. Although the report allowed that "the experts who work on behalf of the Ministry of Economic Affairs are . . . generalists and can't necessarily identify the Angolan gems" by the apparent signatures of the rough, it also underscored that "the importance of the diamond industry to the Belgian economy is a counter-incentive to a rigorous application of the embargo." Moreover, it was particularly important that Belgium and other diamond centers tighten up their borders because, while some rough stones had identifying features, polished ones were virtually untraceable. This meant that the possibility of isolating Angolan products existed for only as long as they remained uncut and unpolished; afterward, they looked just like other good-quality diamonds and were probably mixed up with parcels from various other regions anyway.

A Rough Trade was published at a significant moment. The international community had already been alerted to the issue by the UN Security Council, but Global Witness took the conversation one step further by explicitly explaining the connection between diamonds and UNITA. During its publicity campaign, the NGO reached out to Martin Rapaport, a diamond industry leader and known renegade who, in the 1970s, during the investment boom, had achieved notoriety when he started printing the *Rapaport Diamond Report*, a weekly price list for white diamonds. It was the first publication of its kind, and many insiders opposed the idea of a formalized public guide that effectively wrested control away from the dealers. By 1999, however, Rapaport and his list had become fixtures, and he'd established himself as a respected (if outspoken) voice within the community.

Rapaport, the son of Holocaust survivors, had an instinctive, emotional response to the suggestion that gemstones—his lifelong vocation—might be causing widespread human suffering. He traveled with Global Witness to the diamond-rich West African nation of Sierra Leone, where the political situation echoed the crisis in Angola. It had started in the early 1990s, when a rebel army called the Revolutionary United Front (RUF) crossed the Liberian border and fomented popular dissatisfaction with the existing government. Members of RUF cast themselves in the role of liberators and focused their attention on Sierra Leone's diamond fields, located in the densely populated Kono district, which was surrounded by thick, steamy jungle. Well before RUF arrived, the Sierra Leoneans had developed an understandably complicated relationship with diamonds. They were by far one of the country's most plentiful natural resources—which also included gold, rutile, and iron—but ever since their discovery by British colonizers in the 1930s, the shiny stones seemed to profit everyone but the country's own citizens. RUF gave Sierra Leoneans the impression that it wanted to capture the mines for the benefit of the people.

But that wasn't the case. RUF had set its sights on Sierra Leone's diamonds for only one purpose, which was to fund itself. It needed money and guns—AK-47s were the weapon of choice—and diamonds could ensure a steady supply of both. Once RUF was integrated into Sierra Leone's political landscape, its leaders turned against the citizens, employing violent tactics to suppress opposition. The rebels gutted the population of Kono, killing the men and raping and murdering the women. They kidnapped the boys and used drugs and other brainwashing techniques to transform them into orphan soldiers, creating an army of chillingly loyal children to carry out deadly operations with names like Clean Sweep and No Living Thing. RUF also had a extremely grisly signature inspired, some say, by the president's well-intentioned if feeble appeal that everyone should "join hands" for peace. RUF made a practice of cutting off civilians' hands and other limbs in fast-paced mass amputations, and these disfigured bodies served as walking warnings to anyone who dared to challenge the front.

When Rapaport visited Kono, he saw for himself the devastation that RUF had caused. Displaced amputees were living in refugee camps, separated from their loved ones by distance and death. Kono, once a bustling region, was a ghost town without amenities or infrastructure. Rapaport was appalled and utterly shaken; when he returned home, he threw his support behind Global Witness, prepared to weather blowback from the diamond community. On April 7, 2000, Rapaport published a heartfelt article in his magazine called "Guilt Trip," which was both a stirring piece of first-person journalism and a plea to his colleagues to take an interest in the African wars. "The diamond industry must address the fact that illegal diamonds from Sierra Leone and other war zones are in fact finding their way into the diamond marketplace," he wrote, before making the point that while Forty-Seventh Street dealers were loath to buy suspected stolen diamonds, they were still willing to hide behind a veil of ignorance when it came to conflict stones. Rapaport implored his peers to rally behind the people of Sierra Leone by willingly participating in discussions about regulation and supporting a market where legitimately sourced diamonds would fetch better prices than smuggled ones. His conclusion—that diamond resources should directly benefit and enrich the citizens who mined them—was an important one, especially for an industry that was built upon the strength of colonial powers.

The conflict diamond issue attracted worldwide attention. American congressmen from both sides of the aisle introduced bills that would legislate the way diamonds were brought into the country. In January 2000, the group Partnership Africa Canada (or PAC), another NGO devoted to the role that natural resources, especially minerals, play in violent conflicts, published its own report about Sierra Leone, called *The Heart of the Matter: Sierra Leone, Diamonds and Human Security*. That summer, the diamond industry's leaders—including Martin Rapaport along with several government representatives from the major diamond-producing and -processing countries like South Africa, Namibia, and Belgium—teamed up in Kimberley, South Africa, with activists from Global Witness and PAC to collectively address the problem and strategize better ways to ensure that diamonds from conflict areas stayed out of the legitimate supply chain.

De Beers was there, too. Corinna Gilfillan, who joined the team at Global Witness in 2002 and worked for almost five years on the conflict diamond issue, recalls the panic that swept through the industry with the mounting realization that diamonds could be irrevocably tainted by the damning association with bloodshed. The entire business operated around the goal of keeping prices stable, and now an ideological stain threatened to do longer-lasting damage than the perennial risk of oversupply. What bride wanted a ring on her finger that reminded her of a grief-stricken African father who'd lost his hand? It echoed another recent cautionary tale from the luxury goods sector. "I think a lot of the industry and the governments thought of the fur trade," Gilfillan remembers. "That was the comparison that we heard all of the time when we went to work with industry or governments, that was really the worst fear. That galvanized the diamond industry and governments to want to take action because they were concerned that was the way things would go, [that] there would be a huge backlash." Anyone who publicly resisted change—and certainly there were those individuals who did maintain that it was impossible to track tiny diamonds—risked being cast in the role of international villain.

In July 2000, the newly created World Diamond Council proposed a series of preventive measures that won approval from NGOs including Global Witness and Amnesty International. The new rules, which applied to rough diamonds, specified that stones should be shipped in sealed packages to prevent any mixing or tampering, and that import certificates must list the location of the mine of origin and not just the place of export. This would make it more difficult for smugglers to traffic stones through border countries, the way RUF had been using traders to funnel diamonds through Liberia. These provisions were a good start and showed a willingness on the part of the industry to cooperate, even if it was motivated by the fear of future public relations nightmares. But while the head of Antwerp's High Diamond Council assured the *New York Times* that "any trader that has dealt with these illicit diamonds will be banned out of the business," the international community couldn't simply hand over all responsibility for regulation to the diamond trade; self-policing was admirable, but also not the surest way to guarantee compliance. For the sake of the people of Angola, Sierra Leone, and also the Democratic Republic of the Congo and the Ivory Coast, the world needed a more comprehensive, binding plan for dealing with conflict stones. In December 2000, the UN General Assembly passed a resolution in support of an international certification scheme for rough diamonds.

The next step was to draft it.

◇◇

At the height of the crisis, it was estimated that between 4 percent and 15 percent of diamonds in the supply chain came from conflict areas, with the former number supplied by the diamond industry and the latter calculated by Global Witness. Unfortunately, there was nothing so obvious to distinguish the sparkling clean diamonds from the dirty ones. And so, interested parties like the UN, NGOs, and diamond industry representatives came up with the Kimberley Process Certification Scheme: an international agreement completed in Switzerland in November 2002 and officially established on the first day of January 2003.

The Kimberley Process defined conflict diamonds as rough stones used by rebel movements to undermine legitimate governments. The core regulations simply tightened existing international import/export laws by requiring that all parcels be shipped in tamper-proof containers and arrive accompanied by a valid certificate of origin. But the Kimberley Process took it one step further by stipulating that participating governments could trade only with other Kimberley Process countries, meaning those that had also pledged to meet the minimum requirements. Member countries—which included major diamond-producing regions like Angola, Australia, Botswana, Canada, Namibia, Russia, Sierra Leone, and South Africa, as well as important diamond-trading areas like all of the European Union, India, Israel, and the United States—had an obligation to uphold several practices aimed at monitoring the movement of diamonds across the world. For instance, member states were expected to keep extensive statistical records, subject to audit, for the purpose of rooting out any suspicious anomalies, such as a country with a sharp increase in exports but no known new diamond discoveries.

Corinna Gilfillan of Global Witness describes the long Kimberley Process negotiations as well as the initial implementation as "fraught" with so many different interests involved in the planning: the governments themselves, but also representatives from the diamond industry and the NGOs, like Global Witness and Partnership Africa Canada. "I think one of the real challenges of the scheme . . . was how it was going to be enforced and have teeth," Gilfillan says. Ultimately, the threat built into the Kimberley Process was an economic one, which meant it wouldn't necessarily work. In theory, countries that refused to buy into the agreement, or that flagrantly violated its terms, risked having their diamonds labeled as illegitimate or black market, and therefore risked sacrificing gross revenue. But—and this remains one of the major criticisms of the Kimberley Process—that assumes other participating

governments not only are robustly implementing their own regulations but also are willing to risk their global relationships by
holding other governments accountable.

Major media outlets first broke the news about the Kimberley
Process negotiations in 2001, but it wasn't until a few years later
that the issue of conflict diamonds received widespread attention in
pop culture, which gave rise to a second round of public conscious-
ness and indignation, well after a system had been put into place
to address the problem. The term used by the industry to describe
illicit stones, and the one that appears exclusively on the Kimberley Process documents, is "conflict diamonds"; however, the press
used the macabre synonym "blood diamonds" interchangeably.
In September 2002, two months before word got out that the certification scheme was completed and going into effect, journalist
Greg Campbell published a book called *Blood Diamonds*, which described the violence perpetrated by RUF in Sierra Leone. Campbell
had been on the ground in Kono and corroborated the stories told
by Martin Rapaport and Partnership Africa Canada about the vicious RUF leaders, the boy soldiers, and the amputees. His thesis
also matched the one that had been advanced by the NGOs: Foday
Sankoh, the leader of RUF, had waged his war for control of the diamond fields. "Since the grisly execution of Operation Clean Sweep
in 1996," Campbell wrote, "RUF rebels have sold millions of dollars
worth of Kono's diamonds into the world's marketing channels,
diamonds that are now undoubtedly treasured and adored by husbands and wives who have no idea of their brutal origins."

The book was well reviewed but gained an infinitely larger
reach when it was developed into a big-budget Hollywood movie in
2006. The film starred Leonardo DiCaprio as a fictional, morally
flexible South African diamond trader who experiences a crisis of

conscience after getting to know a Sierra Leonean father, played by Djimon Hounsou, desperately searching for his young son, kidnapped by RUF. The film *Blood Diamond* put the book's most gruesome images into the view of the masses and cast Jennifer Connelly in the role of a fearless investigative journalist, determined not to be confused with the kind of silly American woman who pines for a glittering ring. When the movie premiered, "blood diamond" was already becoming a part of the typical viewer's lexicon, thanks in part to Greg Campbell's book but also to a popular music video by Kanye West for a song called "Diamonds from Sierra Leone." West's original version of the track, which sampled Shirley Bassey's 1971 song "Diamonds Are Forever," actually had very little to do with the West African country; rather, it addressed West's own bad behavior at awards shows, as well as rumors that his label, Roc-A-Fella Records, was in trouble. West admitted that while he might be petulant, the latter couldn't be further from the truth. "Throw your diamonds in the sky," he chanted, referring to the hand gesture made by holding the thumb and index finger together in a diamond shape, which had become the unofficial Roc-A-Fella symbol at concerts.

However, sometime after recording the first version of the song, West became more acutely interested in the plight of Sierra Leoneans and produced a remix, which included additional verses that directly addressed the issue ("These ain't conflict diamonds, is they Jacob?"). In the new lyrics, he positioned the blood diamond problem as an undeniably racial one, comparing the suffering experienced by African Americans at the hands of the drug trade to the anguish many Africans had endured in the name of diamonds. The irony, for West, was that some black Americans like himself had unwittingly enabled violence in places like Sierra Leone not only by buying gemstones but also by promoting a narrative that equated sparkly "ice" with success. The corresponding music video, directed by Hype Williams, opened with chilling testimony from a child soldier, set against images of skinny, inky-eyed kids working with pickaxes in an underground mine. It closed with a caption as West walked into the distance, his back turned to the lens: "Please purchase conflict free diamonds."

"Diamonds from Sierra Leone" arrived at the start of a new microtrend in hip-hop to acknowledge the grim realities of conflict diamonds. While many rappers continued rhyming about their jewelry as if nothing had changed, others like Kanye West participated in a movement to educate their fans as well as their peers. In 2005, Chuck D from Public Enemy narrated an eleven-minute documentary short "Bling: Consequences and Repercussions"; a few years later, *Bling'd: Blood, Diamonds, and Hip Hop* debuted, a VH1 Rock Doc in which Raekwon (of the Wu-Tang Clan), Paul Wall, and Tego Calderón actually travel to Sierra Leone to observe the devastation firsthand. The message was clear: anyone who bought "bling" should know where it came from. Asking the jeweler a few simple questions could, theoretically, relieve important ethical and psychic burdens.

Relative to wealthy white Americans, the predominantly black hip-hop community had been wearing diamonds for a very short time, and yet the racial connection between rappers and victims in countries like Angola and Sierra Leone placed an arguably disproportionate amount of responsibility on socially conscious musicians to get the message out. "It's a very bloody transatlantic connection," Zandile Blay, a fashion journalist and editor in chief of *Hello! Nigeria*, points out. "It's a very sick, ironic, sad, dirty, weird connection that we have. The same black rapper that's position-

ing with his chain has a brother—if you traced the connection—
who could be a relative who got his limbs cut off so he could wear
that diamond. It's just heavy." Hip-hop did plenty to reinforce the
allure of diamonds, but at the same time, as Blay signals, rappers
represented an extremely small portion of the African American
population. The vast majority of black Americans weren't buying
blingy chains, just as typical white consumers weren't shopping
for outrageously expensive pieces like the Taylor-Burton Diamond.
The difference is that the black community hadn't been targeted
by decades of De Beers advertising that equated diamonds with all
manner of personal achievements. Bevy Smith, who is the former
beauty advertising director for *Vibe* magazine, points out: "I don't
think that diamonds have been marketed to black communities.
Tiffany's advertised in *Essence* magazine, [but] I've never seen a diamond industry ad in any black magazine. . . . I will say that they
never called *Vibe* to [express interest] in advertising—that's very
telling."

De Beers never bought space in *Vibe*, yet its stones ended up in
hip-hop photo spreads anyway. Despite the conflict diamond crisis,
the power of this kind of organic, grassroots marketing would only
intensify as time went on.

∞∞∞∞∞∞∞∞∞∞∞∞∞∞∞∞∞∞∞∞∞∞∞∞∞∞

In the years between 1998, when Global Witness published its
initial report, and today, De Beers has dramatically retooled its
business model. This was a question of survival: the conflict diamond crisis reflected terribly on the industry, and De Beers, as the
company responsible for more than half a century of diamond advertising and therefore one of the most recognizable names in the

business, bore the brunt of the public's outrage. It certainly didn't help that the film *Blood Diamond* portrayed the fictional South African "Van de Kaap Diamonds" as a greedy and duplicitous cartel, cooperating with the United Nations while secretly continuing to purchase stones tainted by violence on the open market. But even before then, there were signs that the "good old days" for De Beers, when it controlled almost every aspect of the supply chain from the ground up, were coming to a close. Australia's Argyle mine alone had significantly disrupted the power balance when it opted out of its contract with the syndicate and established its own distribution channels. Also, two mines had gone into production in the frigid northern territories of Canada, south of the Arctic Circle and accessible by seasonal ice roads, both still economically viable because they consistently had such top-notch yields. Neither was owned or operated by De Beers.

These weren't the only challenges—others were political in nature. After the Soviet Union disintegrated, the Russian government went into partnership with a domestic mining company and formed Alrosa, a proprietary diamond business still willing to collaborate with its number one competitor, but only to a point. In South Africa, enormous social changes were afoot. The apartheid government had toppled in the mid-1990s, giving way to Black Economic Empowerment: a mandate for citizens of color that entitled them, among other things, to benefit from the country's natural resources. In Namibia and Botswana, two neighboring African nations where De Beers also had a significant presence, there was increasing pressure from both governments to use diamonds to enrich the communities and advance the interests of their own people, too.

De Beers was losing ground and, with it, the strength of its

once formidable global negotiating position. By the time disturb-
ing stories about Angola and Sierra Leone had made their way into
the news, the syndicate controlled only an estimated 65 percent
of the world's supply of diamonds, down from 80 to 90 percent at
the height of its influence. In 2000, Harry Oppenheimer died at
the age of ninety-one, leaving his son, Nicholas "Nicky" Oppenheimer, as his successor. The younger Oppenheimer took immediate action. In early 2001, he formed a consortium to buy up shares
of De Beers and take the company private. The new owners were the
Oppenheimer family; the Anglo American Corporation (a company
founded by Oppenheimer's grandfather Ernest); and Debswana, a
venture jointly held by De Beers and the government of Botswana.

This was part of a larger overall strategy. A few months later, the
company made a shocking announcement: it was dismantling Cecil
Rhodes's hard-earned monopoly, and was committed to marketing
only those rough stones that it had mined itself. While ostensibly a
reaction to the bad press surrounding conflict diamonds, this news
had other, subtler implications for the industry as a whole. Yes, by
limiting its diamond supply only to the rough that came directly
from its own operations, De Beers could effectively guarantee the
nonviolent provenance of the stones. But the company also had
plans for what to do with those diamonds, and for the first time, De
Beers wasn't interested in forwarding the agenda of the industry
as a collective entity. The new goals were much more specific and
had to do with branding its own products, rather than investing in
the generic diamond marketing that had characterized its advertising since that first campaign in 1939. Instead, it wanted to forge
a tighter relationship with prospective buyers. Research showed
that, increasingly, consumers in the United States, Europe, and
Asia were attracted to branded luxury products, and were willing

to pay a premium for a name and a logo. Even as the syndicate had embraced its historical role as the steward of the diamond world, it had been locked in fierce competition with other mining interests while lending a hand to the jewelers. Now, things would be different: De Beers would increase its downstream activities by selling directly to the consumer, alongside stores like Tiffany & Co. and Harry Winston.

There were some loose ends to tie up first. Ever since World War II, when the syndicate tussled with the Roosevelt administration over industrial diamonds, and the US government retaliated with accusations that the cartel operated in violation of the Sherman Act, De Beers had been prohibited from selling its stones to American consumers. The charges were dropped in 1948, but the relationship had never been mended, and as recently as 1994, De Beers was once again indicted by the US Justice Department, this time over allegations that it had colluded with none other than its former rival General Electric to fix the price of industrial stones. The charges against GE didn't stand up in court—a judge cited insufficient evidence—and De Beers executives refused to settle, convinced that the government's case against them was also too flimsy to take seriously. However, a decade later, in July 2004, they made the decision to reconcile yet another American antitrust suit, this one over the sale of rough. With no more outstanding Sherman Act violations in the States, De Beers—for the first time in almost sixty years—was finally free to do something it had never done before: openly sell to US buyers in its own retail stores.

Today, De Beers controls 30 to 35 percent of the world's diamonds, and even among those stones, it shares ownership with governments of the countries where the actual mines are located. In November 2011, chairman Nicky Oppenheimer sold the family's

40 percent share to Anglo American for roughly $5.1 billion, effectively removing the Oppenheimers' stake in the company. De Beers continues to market its diamonds through a corporation called the Diamond Trading Company, or DTC: a friendlier, less imposing name for the company's rough distribution arm, once called the Central Selling Organization, or CSO. The primary method is still the sightholder system, in which a small number of select businesses are given the advantage of buying boxes of diamonds from De Beers.

However, the DTC has ostensibly been transformed in significant ways from the days of the CSO, too. For one thing, the sights now are held in Botswana instead of England. Since November 2013, sightholders fly to Gaborone six times a year to collect their supply of gemstones for the upcoming season. After almost eighty years of sights in London, the move created some disruption, especially for Americans, who found their flight times more than doubled. "There were some people who were not particularly keen, at least initially," David Johnson, a spokesman for De Beers, recalls, but De Beers's commitment wasn't to the diamond buyers who would be doing the traveling. Rather, it was to the nation of Botswana. Roughly a third of the country's gross domestic product comes from diamonds, and the president of Botswana had advocated for his citizens by effectively pressuring their new business partner to make a move that would have terrific economic implications for them. There had already been improvements: beyond just mining, Botswana was developing a strong manufacturing base, with an estimated four thousand new jobs created in cutting and polishing. But the sights provided opportunities for another level of development: infrastructure, hotels, restaurants, Internet accessibility, wireless. In 2011, when De Beers agreed to relocate the sights, the company had no choice but to invest millions of dollars in Gaborone in order to create a comfortable and appealing environment for its clients.

These days, the sights are more flexible than they were back in the days of Monty Charles, the stern enforcer of the original system. At the start of each twelve-month period, diamantaires submit their requests for stones for the upcoming year and work with De Beers to measure how, exactly, they'd like the year's supply apportioned. When the client shows up in Gaborone, Johnson explains, he or she already has a pretty clear sense of what's going to be in the box. Then, after a coffee, perhaps, and some casual socializing, each sightholder goes into a private room to inspect his or her allocation. The box is divided into smaller packages, and if one or two don't suit the client's needs, he has the option of doing a "buyback," in which he refuses a percentage of the offering and De Beers reevaluates the total price in response.

But while De Beers has become more accommodating with the process of doling out rough diamonds, the criteria for accepting sightholders are increasingly stringent. Johnson estimates that about half of the applicants for the sights actually get to Gaborone, and the ones who are chosen stand out for some combination of a strong market position, good technical capabilities, encouraging financials, and a clear marketing plan. Sightholders are also required to subscribe to a code of conduct called the Best Practice Principles, which stipulates compliance with the Kimberley Process, of course, but also covers other issues like financial transparency, environmental responsibility, and labor standards. There are still real economic benefits to buying rough directly from De Beers; it sources from several mines, and thus can offer a greater variety and consistency of product than other suppliers, creating an incentive, even beyond the obvious ethical concerns, for clients to walk the straight and narrow. However, for the first time in the history of the company, De Beers functions as something beyond just a distributor to its sightholders. With the company's new focus on the world of retail, it has become a clear-cut competitor.

By way of the brand's public face, De Beers has attempted to recast itself and go from the perceived villain to something entirely opposite: a self-appointed positive force of change for the diamond world in the twenty-first century. This message extends all the way down to the marketing plan for its two luxury companies: De Beers Jewellery and Forevermark. The former is a high-end jeweler, which touts its use of superior-quality, ethically sourced diamonds as one of its biggest selling points. The copy on the website is reminiscent of the early days of De Beers advertising; the "Jeweller of Light" calls its diamonds "nature's most exquisite and magnificent prizes." However, when it comes to the De Beers cartel, not everyone is so quick to accept the "cleaner" image at the expense of turning a blind eye to the legacy of darkness. In London in 2002, at the opening of De Beers's first brick-and-mortar store, protesters gathered to bring attention to allegations that the syndicate—with the blessing of the Botswana government—had driven indigenous tribesmen from the Central Kalahari Game Reserve in order to make room for new mines. They carried satirical signs that read "Bushmen aren't forever." While the president of Botswana spoke out in support of De Beers, explaining that relocation was part of a nationwide strategy to provide all of its citizens with access to infrastructure like schools, and had nothing at all to do with diamonds, the shadow cast by De Beers's turbulent history constantly threatens its attempts at a new, squeaky-clean image.

But it keeps trying to burnish the image. The "Forevermark"—a diamond-shaped logo consisting of four symmetrical brush-strokes, referring to Frances Gerety's now classic slogan—became a key image in De Beers's advertising, as part of its major corporate restructuring and rebranding exercise in 2000, in direct response to the conflict diamond issue. The company Forevermark launched afterward, first in Asia in 2008 and then in the United States in 2011. Forevermark—a De Beers affiliate with the benefit of a neutral name—takes an even more straightforward approach than De Beers Jewellery, appealing to those customers for whom the question of a diamond's provenance is their most pressing concern. "Less than 1% of the world's diamonds are eligible" for Forevermark status, the company's written materials claim. That's because Forevermark diamonds are not only good-quality, gem-grade stones; they're also sourced from places where mining benefits the local communities and where environmental considerations are duly met. While one hopes this is true of the diamonds at De Beers Jewellery as well, Forevermark turns the Best Practice Principles into a brand identity. Forevermark stones are actually inscribed, internally, with the corporate logo and an identification number to prove that they qualify. While the inscription isn't visible to the naked eye, it's intended as inalterable proof that the diamond is legitimate and clean.

Whether or not De Beers Jewellery and Forevermark are as popular as places like Tiffany & Co. and Fred Leighton, they serve a very specific purpose beyond just selling diamonds: they communicate a message directly to the consumer that De Beers is concerned about human rights issues and is invested in the health of the planet. Far from imploding on the heels of the conflict diamond crisis, De Beers has managed yet another unlikely recovery by way of a marketing coup, in which it turned global awareness about the problem to its advantage. While Hollywood portrayed the syndicate as a greedy perpetuator of unthinkable violence, De Beers upended its entire corporate structure in order to suit its new identity, as a community leader in the goals of ethical sourcing and transparency. Somehow, the role fitted De Beers pretty well.

A little too well, argue the skeptics.

∞∞∞∞∞∞∞∞∞∞∞∞∞∞∞∞∞∞∞∞∞∞∞∞∞∞∞

In August 2002, the *New York Times* ran a story, "Controversy over Diamonds Made into Virtue by De Beers," which explored this very question: What were the syndicate's motives for aligning with Global Witness? Jack Jolis wishes more people would ask themselves the same thing. Jolis is a rough diamond dealer, expert, and consultant who has been in the business for forty years and is the grandson of Jacob Jolis, who worked as a director for De Beers in London in the 1930s before opening his own company in the States. This company's purpose, Jolis explains, was to help De Beers get around President Roosevelt's antitrust laws by importing De Beers stones through a shell corporation in Bermuda, which proved very lucrative for the Jolis family. In the following years, the family company maintained what Jack Jolis describes as a "like/annoy" relationship with De Beers—sometimes the Jolises were liked, sometimes they were considered annoying. They always knew their current position on the continuum, he jokes, when they surveyed the contents of their boxes at the sights.

Jolis has been irritating plenty of people since the late 1990s, when he claims he threw Charmian Gooch, the founder of Global Witness, out of his office in Belgium. In his view, the entire conflict diamond crisis is a "hoax," a "colossal fraud," and a "manufactured controversy," and he hasn't been shy about saying so, publishing a long letter to the editor in the *National Review* and an op-ed in the *Wall Street Journal*, and even testifying before Congress in 2000, when the Kimberley Process was still under consideration. Jolis knows it's an unpopular opinion, and one that invites accusations of greed and heartlessness. But he speaks out because, as far as he's concerned, the certification scheme is, ironically, having the inverse effect of what it was originally implemented to do. He says it's actually helping De Beers and harming the artisanal diggers in Africa—the very people it was established to protect.

How? First of all, he argues that the whole crisis is a matter of perspective. Jolis doesn't deny that there's been horrific violence in certain regions of Africa, but he vehemently rejects the notion that diamonds, and the people who buy them, are somehow at fault. "The problem with the blood diamond hoax is that, on the surface, it's plausible. It's plausible if you have a certain kind of mind-set which is already mentally disposed not to blame the human agent for mischief but to blame the tools that he uses. In other words, if you're the kind of person who sees a guy sticking up a 7-Eleven with a gun and instead of blaming the guy you blame the gun, that's a certain mind-set," he says, making his conservative politics plain. In his view, the fact that diamond fields happen to be in contested territory doesn't mean that warlike factions are fighting over them specifically, nor does it imply that without diamonds the rebels would be bankrupt. But whether or not he's right—Martin Rapaport and the people at Global Witness would almost certainly disagree— Jolis wants people to know that the Kimberley Process isn't helping anything. To him, it's a gratuitous bureaucratic system that hasn't materially improved the trade; on the contrary, it creates paperwork, facilitates the black market, and has other, more insidious implications, too. Jolis says that it functions to suppress any non-KP-sanctioned diamonds, and this suits De Beers's long-standing strategy to stamp out competitors. Small, independent producers from remote areas are the ones who can't get their stones into the market under the new system, and this, Jolis says, "makes De Beers very happy: fewer non De Beers diamonds out there." By supporting the Kimberley Process, De Beers has officials at the EU and the US customs bureau acting in its service, as its de facto government henchmen.

The certification scheme, he explains, hasn't even effectively done away with diamond smuggling. Rather, it's actually emboldened the underground couriers, who find themselves more in demand than ever before. These illegal traders, who buy diamonds from poor diggers in places like the Central African Republic, where exports are officially *interdites*, are paying less for stones than they used to pay because it's more difficult—and more expensive—to get them across the borders into countries like Cameroon and Chad, from which it's still legal to export gemstones into the United States and the European Union. Meanwhile, the artisanal diggers, who lack resources to do the job themselves, suffer financially. Their lives depend on a simple transaction: selling diamonds for money. But suddenly, the Kimberley Process—an abstraction—is depressing the value of their products.

To Jolis, this is the sad reality of the current system, and one that very few people in the industry want to acknowledge because it's so easy to point to the Kimberley Process as the clear-cut answer to the conflict diamond problem. As long as the KP is in place, the crisis is apparently over, making the scheme invaluable. It provides "moral cover . . . a whole gigantic cover-your-ass operation," he says. And this, surprisingly enough, is a point that Jack Jolis—

who maintains that the blood diamond story is "as bogus as 'a diamond is forever,'" and who is regarded by the industry as a bit of a kook—and the NGOs, who were the first ones to tell that story, can absolutely agree upon.

In December 2011, Global Witness publicly withdrew from the Kimberley Process. Tensions had been escalating for a while, but they came to a head over the majority decision to continue allowing the government of Zimbabwe to export diamonds from the Marange fields. There, the human rights violations resembled the ones that occurred in Angola and Sierra Leone, but in this case they were state-sanctioned, and the diamond fields had been brutally seized by the legitimate government, which maintained its seat of power by intimidating voters. The Kimberley Process had no controls established for this situation—the language of the founding document extended only to stopping rebel factions.

Increasingly, the architects of the KP have turned away from it; for example, Ian Smillie of Partnership Africa Canada and Martin Rapaport have both criticized the scheme for acting as a smoke screen and lulling consumers into a false sense of security about purchasing diamonds. There's a growing advocacy movement devoted to the idea of "fair trade" diamonds, which, just like other products that have earned that label, would originate from safe and environmentally friendly mines and benefit the populations that are actually doing the digging—leading, ideally, to greater prosperity in the developing world. The Diamond Development Initiative, a nonprofit based in Washington, DC, and chaired by Smillie, seeks to introduce a new global partnership, like the Kimberley Process but with broader and more humanitarian goals.

In the meantime, it's the shopper's responsibility to research diamonds and to make the larger social issues a matter of personal importance by doing his or her due diligence before mindlessly plunking down a credit card. "We're not advocating for a ban on diamonds," Corinna Gilfillan at Global Witness says; this may or may not be a comfort to the increasing number of conflicted young men and women who understand that diamonds can be dangerous—nodding to that old rejected Hollywood film title—but also have trouble letting go of deeply internalized messages about the stones being glamorous and romantic. Many twenty-first-century lovers are well informed, but still yearn for the beauty of a diamond and the experience of a time-tested ritual. To those consumers, Gilfillan proposes that they can shop in good conscience as long as they "ask questions and only buy from companies that have clear policies and can give consumers their answers and assurances. How can you show me that you're not sourcing from Zimbabwe, or can you show me where your diamonds are actually coming from? I think that's the fundamental question, and I think more and more consumers are expecting that."

16

The Innovators

HOW TO SELL DIAMONDS IN A (MORE) ENLIGHTENED WORLD

Los Angeles and New York

On February 26, 2012, a young but established actress named Michelle Williams appeared on the red carpet at the Eighty-Fourth Academy Awards. She was nominated for Best Actress that year: the third nomination of her career. Her film was humble compared with the ones competing for Best Picture—*The Artist*, *Moneyball*, *The Help*—but it was also produced and distributed by the Weinstein brothers, known to rack up big wins for their movies. More, it was a biopic, which had given Williams the opportunity to transform herself. And the subject was one of the most famous women in history, herself immortalized over and over again on celluloid, which made the challenge even greater. It took a special actress to channel the charisma and on-screen star power of someone like Marilyn Monroe. Not to mention the voice—Monroe was known for her breathy purr, and it would have been easy for an impersonation to fall flat or to slide into the realm of caricature. But Williams, frequently praised for her moving, restrained performances, didn't stumble in *My Week with Marilyn*. When she arrived at the Oscars, she had already won a Golden Globe for her performance, which made her one to watch in her category, even though she was pitted against acknowledged heavy hitters Meryl Streep, Glenn Close, and Viola Davis.

This, of course, meant the world's greatest couturiers and luxury houses had all angled to have Williams wearing their garments and accessories at the moment she stepped out of her limousine. The practice of dressing celebrities for awards shows had evolved quite a bit since that first ceremony at the Roosevelt Hotel back in 1929. Even in the 1980s and early 1990s, it wasn't at all unusual for a star to choose her own outfit. But over the past twenty years, an entire industry had materialized around the practice of getting celebrities ready for their red-carpet appearances. For an A-list actress, and an ingenue especially, the Oscars are the pinnacle of high-profile fashion events and represent the culminating achievements of her publicists and stylist. What happens in the hours before the show, and then in the days and weeks afterward, can have enormous implications for her career.

It can have incredible impact on a jeweler's or designer's career, too. Now, when a luxury house wants a press opportunity that communicates "glamour," its team will look no further than the nearest red-carpet event, which can all but guarantee a slew of fresh media attention. That night, Michelle Williams stepped out wearing a strapless coral silk chiffon gown with a peplum flounce at the waist, designed by Louis Vuitton. Williams, who wore her blond hair in a flattering pixie cut, bore a closer off-screen resemblance to actresses like Mia Farrow and Jean Seberg than she did to Marilyn Monroe. With fine features and a petite frame, she had a youthful appearance that worked in her favor, especially given the complexity of her performances. But it didn't make sense to put a small woman like Williams in heavy, overpowering jewelry. The best pieces would flatter rather than upstage her, like the timeless Fred Leighton diamond collar necklace that she wore at the Oscars back in 2006 with a mustard-yellow Vera Wang gown and red lipstick. That look—debuted the year she received her first Oscar nod, for Best Supporting Actress in *Brokeback Mountain*—helped her transition from a former star of television teen drama *Dawson's Creek* into a legitimate Hollywood player. It also landed her at the top of almost every "Best Dressed" list, which was great press for everyone involved.

It's no wonder that six years later, when Fred Leighton was invited to accessorize Michelle Williams again, the team members jumped at the chance. They were no strangers to the red carpet and the positive effects it could have for their brand. These days, Fred Leighton rivals the finest high jewelers in the world, but the house's origins are actually quite humble. It started out as an off-the-beaten-path Greenwich Village boutique in the 1960s, when a New York native named Murray Mondschein, a taxi driver's son, purchased a third world imports store of that name. Mondschein was an ex-army man with a softer side; he had an interest in folk fashion and wanted a place to sell his unusual treasures, namely crafts and lace dresses from Mexico. Soon, he was importing special fabrics and using them to create garments of his own design. Among the assorted bric-a-brac he carried, there was an increasing amount of jewelry, which he, a born salesman who excelled at chatting up his customers, easily persuaded them to buy along with their new dresses. Eventually, vintage and estate jewelry took on a leading role in the business, and Mondschein moved his store to the Upper East Side. By then, he'd assumed an alternative moniker: Fred Leighton. People who called and stopped in always wanted to speak to the man whose name was on the awning. Tired of explaining that he'd inherited it from a previous owner, Mondschein took the path of least resistance and started answering to Fred.

He began collecting pieces from a golden age of jewelry: baubles from the 1930s, '40s, and '50s, created by houses like Bulgari, Chopard, Van Cleef & Arpels, and Cartier and, in some cases, designed by big-name artists like Jeanne Toussaint and Jean Schlumberger. At the time, vintage jewelry wasn't terribly popular, and Mondschein paid modest prices for his finds, hauling them back to New York and selling them to well-heeled patrons with the promise that one day signed pieces—ones that not only were beautiful but also had a distinctive pedigree—would increase tremendously in value. He was absolutely right. By the early 1990s, when Fred Leighton first started working with celebrities, the company was known as one of the premier American sources of heritage jewels.

Rebecca Selva, Fred Leighton's chief creative officer and PR director, has been with the business for over twenty years, and remembers how different things were before the red carpet became an industry in its own right: "I came here in 1992, and while we lent for fashion shoots, we did not lend to celebrities. In fact, we had a request from somebody who was getting married, she was very famous and she wanted a tiara, and we were, like, absolutely not! This is out of the question. We don't lend tiaras." But then the grande dame of Italian high fashion, Miuccia Prada, reached out with an unexpected request, and the team at Fred Leighton decided to take a chance.

It was 1996, and she was dressing Nicole Kidman—still married to Tom Cruise at the time—in a simple lavender empire-waist gown for the upcoming Oscars ceremony. Prada envisioned the statuesque actress wearing simple opal jewelry—a surprising choice given that diamonds were still de rigueur for black-tie events and the opal is a significantly less precious stone. After some consideration, the group at Fred Leighton agreed to lend her a unique multistrand opal choker necklace, which was especially flattering on Kidman given her swan-like neck, the simplicity of the dress, and the fact that she wore her wild hair pulled back in a sleek low bun. The evening that she stepped out onto the carpet was a thrilling moment for everyone who dressed her, but it didn't change Hollywood overnight. Selva remembers that she and her teammates celebrated their little triumph, but they didn't even send out a press release—it just wasn't how things were done. Those were the days before the omnipresence of E! and the endless "Best Dressed" lists; rather, Selva would simply get a call here and there from a designer or stylist asking for her help in completing a head-to-toe look. "There really weren't that many people involved [in celebrity styling]," she recalls. "I think everybody did it just to [have fun]—we all knew it was great press, but nobody imagined what it would generate, the industry that would develop."

She cites another red-carpet highlight, also starring Nicole Kidman, as the moment when people started paying more attention to what the stars were wearing and noting the impact an awards show could have on a designer's career. The following year, in 1997, Kidman arrived at the Academy Awards in an embroidered yellow satin gown designed by John Galliano during his first breakout year at Dior. Ever since he had been hired, skeptics had questioned whether Galliano was the right choice for the illustrious house, and Kidman's breathtaking gown was his showstopping answer to all of the criticism. That night, Kidman wore it with a pair of Fred Leighton beaded chandelier earrings and two brightly colored Indian bangle bracelets made of enamel and yellow gold and studded with round precious stones. This jewelry—as a natural follow-up to the opal necklace she had worn the previous year—was remarkable not only because it was gorgeous, but also because it was so different from what the stars around her were modeling. Other actresses wore pieces that eclipsed them: huge gemstones that upstaged the women themselves. "They looked like they were borrowing jewelry," Selva notes. "And all of a sudden you have Nicole Kidman walking out . . . with Indian jewelry and these earrings, and it just made people turn their heads and say, wow, she looks amazing. But she also still looks like Nicole Kidman."

By the time the 2012 Oscars came around, Fred Leighton was an established presence on the red carpet, dressing the likes of Sarah Jessica Parker, Charlize Theron, and Madonna. Over time, the team realized that when a Fred Leighton bauble showed up at one of these coveted events, it did more than just invite good press; it actually had a financial bearing, too: lending an expensive piece to a famous woman often meant that it sold to a noncelebrity customer in the weeks that immediately followed. Awards shows consistently get people in the door; after Claire Danes wore a gold wrist cuff with a minimalist pink gown to the 2011 Golden Globes, a customer came in off the street requesting the bracelet. She knew, from Danes's on-camera interviews, that it was from Fred Leighton. The red carpet has become another venue for product placement, pure and simple—almost exactly the way that De Beers, N. W. Ayer & Son, Paul Flato, and Harry Winston had once envisioned when they began courting Hollywood in the post–World War II era. It has become so essential to the luxury business that many designers now pay celebrities to model products at their upcoming press appearances.

It was actually the people at De Beers who initiated the conversation with Fred Leighton about dressing Michelle Williams for the 2012 Oscars, by way of their budding retail brand, Forevermark. To drum up publicity, Forevermark made a practice of dressing red-carpet stars, starting with—who else?—Nicole Kidman at the Academy Awards in 2008. There, the statuesque redhead stunned in a simple black gown and an enormous, multistrand white diamond sautoir designed by L'Wren Scott and made up of 7,645 diamonds dripping from the necklace in strands like tiny icicles.

De Beers offered Fred Leighton a chance to collaborate by creating new pieces for Michelle Williams using Forevermark proprietary diamonds. A handful of jewels resulted, but only two made it onto the carpet with Williams's Louis Vuitton gown: an elegant rivière necklace, consisting of 10.69 carats' worth of sizable matched stones, and an eleven-carat antique bow brooch with a Forevermark diamond in the center, which Williams wore as an asymmetrical detail at her waist. The overall look was elegant: the perfect mix of polished and effortless dressing. Williams, unfortunately, didn't take home a statue that night, but her red-carpet appearance received almost universal praise, which constituted a different kind of win, and not just for her; it was a win for Louis Vuitton, a win for Fred Leighton, and a win for Forevermark.

◇◇◇◇◇◇◇◇◇◇◇◇◇◇◇◇◇◇◇◇◇◇◇◇◇◇

The rise of the red carpet in the shadow of the conflict diamond issue shows that even the bad press about blood diamonds couldn't drown out the other, by now classic, associations with gemstones: the ideas of glamour, romance, status, and love. Rather, the new politically and morally charged information got filtered into the existing brew, producing a slightly bitter aftertaste, so to speak, but without fundamentally altering the flavor. Greg Kwiat—whose family owns the century-old Kwiat jewelers as well as its more recent acquisition, Fred Leighton, which it bought from Murray Mondschein when he retired—describes how increased awareness about diamond sourcing has affected the conversation at the point of sale: "I think it matters to customers at a certain level. I think that once they're comfortable with whom they're working they focus then on the next and, as I would describe it, more important elements of how they're feeling about the jewelry and what they want in the piece itself."

And what they want, more often than not, is the very same thing that their mothers and grandmothers wanted: a shiny diamond to communicate marital status and to symbolize depth of love. In a 2013 annual survey of engagement and wedding trends, popular website *The Knot* found that 86 percent of grooms proposed with a ring, paying an average of $5,598—up from $5,431 in the previous year—for a piece of jewelry with a mean 1.1-carat center stone. The survey of thirteen thousand brides and grooms found that more and more couples are shopping together and that many women browse for rings online before settling on a particular style. But whether the couple ultimately makes a purchase from an affordable mass-market store or from an upscale heritage brand, one thing is absolutely certain: the diamond engagement ring, shortened to "DER" in industry memos, is still by far the most significant bauble from a seller's perspective. "It's an important part of the strategy of our company," Greg Kwiat confirms, "and the reason is, from the jeweler's perspective, meeting somebody at the moment of an engagement is a great way to introduce them to your company and your collection, and if they love what you do, and they loved the engagement ring, you've made a loyal friend and client for life." It's the same thing Cecil Rhodes suspected when he consolidated the industry over one hundred years ago. The memory of an engagement is indelible, and when it's connected with a diamond, the light of that moment flickers forever.

Today, the ritual of giving and receiving an engagement ring feels, simultaneously, profoundly traditional and yet also cutting-edge. It's become like any other genuine milestone: personal but almost universal, too. Television commercials for diamonds highlight the repetitive nature of the classic proposal. Whether or not it's intentional on the part of the advertisers, these fifteen-second or thirty-second spots always hew tightly to the same basic formula: a man kneels, produces a box, and the woman goes wide-eyed, then cries happy tears. There's a kiss, and then he slips the ring onto her finger. Audiences know this scene fluently, and yet the industry spends money year after year to reinforce it rather than innovate beyond it. Which ultimately raises the question: Is the narrative always the same because that's how it happens in real life? Or maybe—and this is where N. W. Ayer & Son gets a lot of the credit, even though De Beers let Ayer go for good in 1995—it happens that way because the diamond industry, through countless advertisements, pamphlets, and Hollywood films, has patiently and systematically taught us exactly how a proposal should look, exactly how we should behave.

But there have also been recent developments in the ongoing conversation about engagement rings that De Beers can't take credit for. Beyoncé's anthemic "Single Ladies" is a modern-day

"Diamonds Are a Girl's Best Friend," with an added dose of empowerment and sass. The ring is still positioned as a trophy; it follows then that the single ladies who overnight become affianced celebrate by posting pictures of their freshly jeweled left hands on websites like Facebook and Instagram. Social media universalizes the image of an engagement ring so that girls and women of different ages, ethnicities, and economic backgrounds can all see just what happens immediately after a couple agrees to get married. The "ring selfie" trend wasn't the work of the diamond industry, but it's certainly welcome, serving the role that traveling "Diamond Lady" Gladys B. Hannaford once had when she showed up at schools to instruct students in the finer points of a time-honored tradition. Now, sixty years later, young girls are indoctrinated by their own role models rather than by a charming, bespectacled matron. The diamond goes along with other achievements: success, but also domestic bliss. It's the "personal Olympics," Zandile Blay quips. But unlike Olympic gold, the ring is a medal anyone can win.

∞∞∞∞∞∞∞∞∞∞∞∞∞∞∞∞∞∞∞∞∞∞∞∞∞∞∞

The jewelry industry has changed tremendously in recent years, and not just because the conflict diamond issue has ushered in new regulations for international trading. Rather, different priorities for consumers, along with different imperatives for sellers, have together altered the landscape of the diamond world. When it comes to engagement rings, for instance, more and more couples are shopping for rings together, as jewelers can attest. The ring is still extremely important for most pairs, but the element of surprise—once a cornerstone of the marriage proposal—is not nearly as crucial. "In general people are already living together, and have been together for a long time, and I think they're just used to making decisions together, especially important ones. They might even already have joint finances," explains Elizabeth Doyle, who owns a New York City vintage and antique jewelry store called Doyle & Doyle with her sister. Doyle has noticed a trend of women coming into her store to "preshop," meaning that they pick out an assortment of rings and then the man sweeps in for the final decision. Maybe that's how modern-day romantics get to have it both ways: the woman gets a voice, but the traditional gesture of the man presenting a ring box is preserved.

For many contemporary brides, the sense that her ring is unique is essential, too. Fewer women are opting for a simple Tiffany Setting, in part because when everyone wears the same style, the only point of comparison is the carat weight of the center stone, which is still part of the fun for some women, while others, increasingly, admit that it makes them uncomfortable. Sometimes this means that couples look for a vintage or antique ring: a true one-off. Others who prefer the modern look gravitate toward vendors that make customization easy. This helps to explain the success of an Internet retailer like Blue Nile, which is unusual because it has no brick-and-mortar counterpart. Couples who use the site sacrifice the immediate and tactile experience of handling the product before they purchase, but they're also getting specific advantages that buyers once enjoyed long ago when ordering from the Sears, Roebuck catalogue. Online, customers can choose everything, including the size, shape, and quality of their stone, and then have it fitted in a setting that best suits their taste and budget. Shopping for diamonds in this way isn't necessarily glamorous—it's more akin to ordering takeout Chinese for dinner than taking a trip to nineteenth-century Paris—but it's appealing for other reasons. It's

affordable and puts the customer back in the driver's seat. For instance, there's no pushy sales representative standing at attention
and hoping to balloon the budget, reminding the couple of the two
months' salary guideline.

And at the other end of the spectrum, the storied Fifth Avenue
jewelers have been transformed, too. Tiffany & Co. is now publicly
traded; Cartier and Van Cleef & Arpels are both owned by the massive luxury group Richemont. Bulgari is a holding in LVMH's portfolio, while the Dominion Diamond Corporation, a mining company with a large majority stake in one prolific Canadian mine and
a 40 percent share in another, purchased Harry Winston from his
two sons. It means that all of these originally homegrown jewelers now have priorities well beyond making the finest baubles in
the world and burnishing the family name. Average shoppers are
more likely to come across Cartier perfume than Cartier jewels;
chain department stores carry everything from Bulgari sunglasses
to bath soaps. Within the luxury goods market, diversification
has generated enormous profits, but some skeptics argue that the
overall cost to the industry has been substantial. Andrea Buccellati, who is the third generation of the Buccellati family to serve as
the house's president, speaks candidly about the effect that parent
companies have had on the best jewelers in the world: "They were
once controlled by a family, or the family had a really strong presence in the company, they had a really strong tradition, you like it or
you don't, but they kept the identity of the style. At the moment the
big group comes in, of course they have a different interest than the
family. The family, usually they don't really—I don't want to say they
don't care about money, but they care more about keeping the company integrated, to grow with the name. . . . They want to keep the
prestige of the company. Groups, they're looking for that, too, but

they always look for more money. At the end of the day, they have to answer to the investor. And this sometimes . . . they kill the image, the heart of the company." Of course, Buccellati is not immune to the economic pressures of maintaining a specialty high-end business; in 2013, he brought on a majority investor, although the family remains on board, active in both the brand's overall vision and its design.

Critics argue that luxury groups have made the world's most famous jewelers homogeneous, where once they all had their own distinct personalities. Douglas Kazanjian, CEO of Kazanjian Bros., a family-owned fourth-generation company based in Beverly Hills that specializes in estate jewelry, suggests that this problem of homogeneity extends even further down through the industry to its most basic level, to the way diamonds are being cut. In January 2006, the GIA introduced a new cut-grading system that categorized the cut quality of all round brilliant white diamonds by measuring how effectively they dispersed light. This, in Kazanjian's view, has oversimplified an extremely complex process by shackling art to science. Like some other experts who primarily trade in older stones, he believes that the antique cuts are actually more beautiful than the modern ones, and the effort to capture and quantify anything as abstract as beauty is completely misguided. The existing fashion, reinforced by GIA, is to cut diamonds so that they shoot out blazes of white light, in the hope of achieving "that icy, blingy whiteness," in Elizabeth Doyle's words. Antique diamonds are warmer, with more rainbow scintillation. Says Kazanjian: "If you take an antique cut diamond today and compare it to the more modern excellent cut stones, the antique cut will actually look more beautiful in the evening than the modern stone. You will see it more easily from across the room, and it will shine brighter.

The light will dance off of the stone in a much more attractive way than it will off of a modern stone, which has a lot of tiny facets trying to shoot as much light back as possible. This is where [GIA has] gone totally wrong. 'Cause they're saying that these stones are better than the antique cut stones."

Ultimately, it's a matter of opinion whether vintage stones are lovelier than the contemporary ones, and families who came up generation after generation in the diamond business are bound to harbor passionate feelings on the subject. But Kazanjian poses a question that has resounding implications for the gemstone world going forward; he asks, when you are choosing a diamond, "Do you want it to look beautiful in a laboratory, or do you want it to look beautiful at night?"

<hr>

These days, the mere mention of a scientific laboratory has the potential to ruffle feathers within the diamond world, and not only for the reasons that Douglas Kazanjian mentions. Since 1970, when General Electric successfully grew its first jewelry-grade gem at the rate of one carat per week, the possibility that scientists would eventually figure out how to more efficiently produce beautiful stones has vexed the industry. Now, in the twenty-first century, that fear has become a reality; consumers can purchase chemically and optically identical diamonds created in labs—or, as those who are involved in the process like to say, "aboveground mines." These stones aren't simulants like Wellingtons, or even cubic zirconias. But—and this is a question that plagues the jewelry establishment— should they really be called "diamonds"?

The debate that has divided the industry actually has a fair amount in common with conversations about the human soul. It's like this: if scientists ever successfully create cyborgs who are virtually indistinguishable from humans and who, in almost every case, can easily serve as charming alternatives, well, suffice it to say, this will invite a good amount of existential panic. As dramatic as it sounds, the anxiety around lab-grown diamonds is similar. For a long time now, since the era of Project Superpressure, those who work in gemstones have been willing to accept that when it comes to industrials, man-made stones and mined ones are equivalent. However, that same logic doesn't extend to jewelry-grade diamonds. There's a feeling that lab-growns not only impinge on the market but also jeopardize the very thing that makes the stones special: the myth and, therefore, their very essence.

Longtime diamond industry executive Lisa Bissell took over a lab-grown diamond business called Gemesis in May 2014, and has been working hard to combat the existing prejudice against her products ever since. Gemesis was the first company to market manufactured diamonds directly to buyers, and had experienced some success, as well as considerable backlash from the traditional diamond world. The business came about when a former US Army general named Carter Clarke, who was enjoying a second-act career as CEO of a department store security company, traveled to Moscow in 1995. When Clarke arrived in Russia, he was in the market for a new kind of ink that was invisible to the naked eye but would reliably set off alarms if it wasn't removed before leaving the store. But his meeting took off in a surprising direction when his contact in Moscow asked him if he had any interest in making gemstones. Clarke found himself taking a long car ride away from the city to check out a diamond-making machine on sale for $57,000 (a little over $88,000 today). He was impressed, but declined to buy it. However, after he returned to New York and still couldn't get the image of Russian lab-grown diamonds out of his mind, he got in touch—and founded Gemesis, out of a warehouse in Sarasota, Florida.

It took almost a decade for Gemesis to get off the ground. The company marketed its man-made, yellow-tinged stones as jewelry; the color was a result of nitrogen impurities, and luckily for Carter Clarke, champagne and canary diamonds were still popular. In order to compete with the established polished diamond market, Gemesis offered its gems at a discounted price of about 50 percent less than traditional mined stones. But its products were so utterly convincing that when six hundred lab-growns in an unmarked parcel somehow made their way into the supply chain in Antwerp, De Beers all but outright accused the Florida manufacturer of trying to pass off the stones as mined diamonds. An incident of this nature was exactly what the mined diamond industry had warned against, and it provided fresh fodder for panic. Gemesis, of course, denied having any involvement, which—as some insiders pointed out—didn't mean that someone with access to its stones wasn't responsible.

Even though the so-called counterfeit diamonds were ultimately discovered, stories like this one stoke fires of anger and doubt within the jewelry world because they have the potential to undermine the perceived value and integrity of the product. Over the past ten years, the luxury category has become increasingly crowded, and as David Johnson from De Beers puts it, "If consumers don't feel confident that when they buy something it is what they think it is, then they'll more than likely move on to another category and buy something that they do feel confident about." Nevertheless, Lisa Bissell believes that there's room for lab-grown diamonds in the jewelry sector. Like Helen Ver Standig before her, she isn't interested in tricking anyone into buying her stones. On the contrary, Bissell is committed to one thing that her colleagues in the mining world keep crying out for: disclosure.

The very first thing Bissell did in her new position was to change the name of the company from Gemesis to Pure Grown Diamonds. Maybe her decision had a little something to do with the recent controversy associated with Gemesis, as well as the fact that the corporate structure had been entirely retooled. Carter Clarke and his Florida manufacturing facility were out. Bissell's Pure Grown Diamonds is based in New York, and it markets and distributes stones that it purchases from a Singapore lab called IIa Technologies. But as Bissell acknowledges, the name Gemesis has other limitations, too: it doesn't tell the buyer a lot about the product. Pure Grown Diamonds is a "snapshot," she says, and puts the word "grown" front and center—which, again, is consistent with her marketing vision for the brand.

Grown diamonds are still a new classification of jewel, and so Pure Grown Diamonds, as the face of an aboveground mine, has done what De Beers and Rio Tinto did so successfully: extensive market research. The findings were positive, and quite clear. The biggest impediment to selling man-made stones was a lack of education about them; people confused lab-growns with simulants like CZ and moissanite (another popular diamond substitute), and this confusion made buyers regard them as inferior to the real thing. Once consumers understood that aboveground mines produce chemically identical diamonds, their opinions shifted.

IIa Technologies is named for the highest classification of stones its labs create: Type IIa diamonds, which are free from impurities like nitrogen and represent less than 2 percent of the ones that occur in nature. In layman's terms: they're usually very, very white. Bissell says that Pure Grown Diamonds gets the best gemstones that IIa Technologies grows and that the aboveground mine's output is not all that different from what comes out of the earth. Mixed qualities emerge from the "greenhouse," Bissell explains. "It's not like a popcorn machine where you put the kernels in and you get pretty much a consistent yield. You start the process and nature takes over, and then you *ooh* and *aah* about what comes out at the end."

The most beautiful gem-grade stones are cut and polished in the same places where mined diamonds travel, mostly in India and China. Then they're set in jewelry and sold to customers—at a price about 25 percent less than natural diamonds of comparable quality. The lower cost is a real selling point, as is the fact that a Singapore laboratory has significantly less potential than a conventional mine to do harm to the earth. The website for IIa Technologies touts the sustainability of its products, pointing out that digging for any kind of precious metals can disrupt the earth's ecology and displace indigenous populations. It also goes without saying that lab-grown diamonds are conflict-free. Bissell, who in her off-hours happens to be a bee conservationist and feels strongly about the environmental consequences of human action, sees great potential for building a customer base with millennials—who are also her target market because they're the age group that is poised to get engaged and buy diamond rings. She says: "It's important [to the millennials] that people are not having their hands and feet chopped off, and [they understand that] if you destroy the earth, you're not going to have an earth . . . and if you say [the diamonds] came from a facility that's really clean, and people are employed, and nobody was hurt, and the earth wasn't damaged, they're, like, right on!" Still, she maintains that these benefits—as appealing as they are—are just one "feature" of lab-grown stones, and the cost-to-quality ratio is actually the major advantage.

It might be just a "feature," but it's a big one; selling a diamond that's guaranteed to be clean is a big deal at a moment when the messaging about mined stones is still shadowed by the term "blood diamond," and industry heavyweights and NGOs alike are deeply critical of the Kimberley Process and its effectiveness. And this, perhaps, is one of the real reasons why the industry finds lab-growns so annoying: they simultaneously undermine the carefully crafted diamond narrative—that diamonds are magical treasures from the earth and incredibly rare—while eluding the other, more onerous story about conflict diamonds, which stubbornly persists despite the industry's long-standing attempts to get the conversation under their control. Meanwhile, the debate about what to call lab-growns, and how to distinguish them from mined diamonds, continues. The industry rallies behind the term "synthetic diamonds," which the still nascent lab-grown business resists—arguing that customers associate the word "synthetic" with something fake, making the term more appropriate for simulants like cubic zirconia. Those with a stake in aboveground mines prefer "created" or "cultured" diamonds, borrowing from the accepted nomenclature for man-made Mikimoto pearls. But establishment diamond people contend that these words don't do enough to signal a difference between naturally occurring diamonds and ones that are scientifically cultivated. The current FTC guidelines require all lab-grown stones to be appropriately and specifically labeled, but manufacturers and distributors aren't required to bill them as "synthetic." However, they aren't allowed to use "cultured" without other modifiers, like "man-made" or "lab-grown," either.

The increased availability of lab-grown diamonds means that there are at least three classifications of colorless stones: simulants (like moissanite and cubic zirconia); man-made diamonds; and natural mined diamonds, which are still the most expensive. Blurring the boundaries between the latter two, many treatments can now be performed to enhance the quality of flawed diamonds. These are extensive: off-color stones might be subjected to irradiation or heat treatment to improve their overall look, and splotchy diamonds might be laser drilled or injected with bleach or acid to remove inclusions. Cracks and feathers can be filled in with glass or silicone, to make the diamond appear more uniform and stable. None of this is necessarily a problem—it's akin to plastic surgery, but for a gem—except that, again, sellers are required to disclose whether or not a stone has been subjected to human intervention, beyond just the typical fashioning. However, it doesn't always happen, either because the seller himself is dishonest or because he doesn't actually know that he's dealing with a treated diamond that at some point got slipped into the pipeline. It's important, not only because unaltered gems are worth more, but also because ones that have undergone an "operation" are sometimes less structurally sound as a result.

Grading labs like GIA work hard to detect any modifications, and in the case of lab-grown diamonds, there are tools available that can identify the differences; some of these tools were developed by De Beers. Pure Grown Diamonds internally inscribes all of its stones with the words "lab-grown" or the letters "LG" along with a certification number, but skeptics still wonder if crooks will be tempted to grind off these markings and sell the stones as mined diamonds for a profit. About this, Bissell admits: "It could happen. There are crazy people out there. . . . Hopefully they'll go to jail for that." But just because that possibility exists, does it mean that man-made gems shouldn't be on the market? As long as the technology continues to improve, there will always be a threat that lab-growns will contaminate the global supply of mined diamonds. But it also means that the world can have a sustainable source of glittering white gems in perpetuity, without harming the earth or the people who inhabit it. This, as it turns out, is worth considering at a time when experts say that the earth's generous bounty of mined stones, which is finite, might be running low.

∞∞∞∞∞∞∞∞∞∞∞∞∞∞∞∞∞∞∞∞∞∞∞

The industry's determination to classify and isolate different types of diamonds boils down to two separate but related imperatives: to protect the commercial value of the product and to preserve the idea that the stones are inherently special. But it's clear that this latter notion is still alive and well in places where access to wealth is just now becoming more widespread, like China and India, and where the emerging middle classes are enjoying newfound prosperity. As a result, China and India are no longer simply home to thriving diamond-cutting centers; they are hosts to an entirely new market of diamond consumers, too. Douglas Kazanjian, whose business specializes in estate jewelry, sees a kind of poetry in this development: "Jewelry travels the world based on wealth. A few hundred years ago, India had the greatest jewelry collections. They loved jewelry more than anyone, they were traveling to Colombia to buy emeralds, trading spices, they had all the great stones, they had the first big diamonds, the Golconda diamonds that are the greatest in the world, the purest color. Then the Europeans started coming into India and were buying their gems . . . and Europe had the greatest jewelry from 1900 until World War II. After World War II, all the jewelry started coming to the United States as the United States became so strong. And now it's going back to China and India again."

The up-and-coming Asian buyers seem to have absorbed what Europeans and Americans have believed for some time: diamonds are not just pretty, shiny things, but also symbols of love, desire, and affluence. And this, above all, is what keeps the market flowing; the concept that David Johnson at De Beers refers to as the "diamond dream" is what makes the stones, as the quintessential luxury products, so untouchable. But the diamond dream is also a distinctly twentieth-century story. Though it certainly relies on images from history—Indian maharajas wearing hulking rough stones and European empresses dripping in their finest Bapst pieces—it thrives on more recent and, therefore, arguably more potent concepts, like middle-class aspiration and Hollywood glamour (which is arguably just another expression of status anxiety). Still, the world continues to change, and in these early years of the twenty-first century, there are questions about how the diamond industry will respond—whether the time-tested messaging will continue to be effective, or if it will have to evolve and adapt. Simulants and lab-growns and conflict diamonds are only a few different facets of a looming and still somewhat shadowy threat.

As of now, however, there's reason to believe that the idea of the diamond dream remains vivid. In 2013, in its third annual report about the diamond industry, the global management consulting firm Bain & Company expressed findings that demand for rough and polished stones would continue to rise through the year 2023, thanks to emerging markets like India and China and, of course, continuing interest from the United States. The report, subtitled *Journey Through the Value Chain*, estimated that over the next ten

years, the number of American couples who get engaged with a diamond ring would hold steady at around 80 percent. More, Bain &
Company observed that the Western ring tradition was catching on
in India. In China, which in 2011 surpassed Japan as the world's
second most voracious diamond consumer, the increased urbanization and meteoric expansion of the middle class—from 19 percent of the population to an expected 44 percent by 2023—mean
that there will be an ever-increasing group of sophisticated consumers with disposable income to spend on things like gemstones.

That wasn't the only good news for the industry. The report also
stated that the market would stay in balance until the tipping point
in 2018: the moment when demand would begin to outpace supply,
thereby driving up the global price of diamonds. If the landscape
of mining remains all but unchanged over the next decade—
accounting for new projects that are already under way in Canada,
Russia, and India—then, as Bain & Company predicts, "the global
supply of rough diamonds is expected to peak in 2018 and decline
as existing mines are depleted." What this means is that, for the
first time since the 1890s, the glittering gems might legitimately
become rare, and not on account of some Cecil Rhodes—esque artificial scarcity, either. From the moment that De Beers took control
of the mines in South Africa, the so-called intrinsic value of a diamond was bolstered by the illusion that it was very hard to come by.
But what if that were suddenly true? For a time, the sellers would
likely reap record profits as a self-fulfilling prophecy finally came
to fruition.

Yet ironically, the industry would be shortsighted to wish for
this, because the end of the trade as we know it is inscribed in this
hypothetical version of events. Lisa Bissell, whose company actually stands to benefit if the global supply of mined diamonds dries

up, insists that she's not waiting for that day. On the contrary, she says, "If it happens, then that's a great loss for everybody because that's part of the economy. You know, say you're Botswana and that's your economy. That's tragic. What does that mean, people can't make a living and feed their family? I don't want to live off somebody's tragedies. But it's a reality. And it's getting worse. It's a mine, it's a resource. And resources are finite." De Beers agrees, at least in part—it's actively prospecting for new mines. However, Liz Chatelain of MVI Marketing, who helped the Australians brand champagne diamonds, points out that even if new kimberlite pipes are discovered, it could be tricky to get them up and running: "I think there's going to be a shortage of natural diamonds. There just aren't that many green fields . . . and wherever you find a mine, I guarantee you it's going to be in a sensitive part of the world."

It's impossible to say how a long-term shortage of diamonds would affect how they're perceived—if, for instance, the sparkling stones will one day be regarded as a vestige of another, embarrassingly extravagant time or if, on the other hand, they will become even more cherished for their uncommonness, as they were in Golconda and the courts of Europe. Perhaps the gap between lab-growns and mined diamonds will expand even further, with lab-growns considered gems for the masses and natural ones reserved for the highest elite. Maybe the mining companies will step aside, admitting defeat, and the market of the future will be less structured, controlled by a network of dealers and estate jewelers trading in an elusive product, purchased by wealthy collectors and connoisseurs. Once again, depending on the stability of the global economy, and assuming the stones hold their value, rich and poor alike might gravitate to diamonds for their potential as a universal and portable currency.

Only time will tell. Meanwhile, there's something about the crystalline, sparkly stones that sticks with us, bound by some tantalizing mix of aspiration, appreciation, and avarice and compelling us to pay exorbitant prices for the privilege of wearing them on our bodies. "A Diamond Is Forever," De Beers continues to assert, and whether or not that's any kind of wisdom, it is true that, for some people, looking into the eye of a polished diamond is like watching the earth and the artisan work together to perform a stunning magic trick. And maybe that's what separates it from other beautiful things the planet has to offer—the way a diamond represents a unique collaboration between nature, sublimely mysterious and unknowable, and human beings, who have somehow found a way to dig up the stones, fashion them so precisely, and even position them so they represent the most abstract desires, like admiration and love. Today, a classic white diamond is a series of contradictions: it's simultaneously brimming with complex meanings while being colorless, empty, and pure as the carbon that gives it structure. For many people, the tiny stone suggests everything from blood and violence to romance and exciting possibilities. The diamond is old and it's new; it's worthless and it's enormously expensive; it's life and it's death.

And perhaps it's those very contradictions that provide us with our biggest clue to the secret of the stone's long-lasting appeal. A diamond is a perfect empty vessel. Like the *Mona Lisa*, a gorgeously cut and polished jewel is both mesmerizing and entirely unknowable—and in many ways this makes it the ideal commercial product. It has no practical function that we, as consumers, will eventually replace or outgrow; it has no intrinsic message that we'll eventually come to reject. This absence of meaning is, paradoxically, what makes diamonds so intensely meaningful, because they lend themselves so well to human projection and—as the people of De Beers, N. W. Ayer & Son, and J. Walter Thompson quickly learned—to the persuasive powers of marketing.

In this context, it makes sense that the gemstones inspired the earliest examples of product placement—they simply sit and wink in the sunlight, inviting onlookers to imagine what, exactly, they're communicating. In places where diamonds are cherished, they've proved to be so resilient precisely because they do and say next to nothing. They've remained quintessential symbols of status and romance even as our notions of status and romance have changed with the times. The greatest asset of the diamond as a tool is most certainly its hardness, but for the diamond as a product, it's something else: mutability.

A diamond is a luminous crystal ball, in which the forecasted future isn't objective but rather a manifestation of our own needs and deepest desires: the image we want to see. It's the spectacular outfit that we buy just in case a glorious evening arises; it's the quick, passing eye contact with an attractive stranger that leaves us happily buzzing with a thousand what-ifs. That's what makes the little stone so potent: it's beautiful, and that beauty stimulates our imaginations and lets us bask in a prism of possibilities. No matter what we learn to be true about the stone, we all, deep down, like getting lost in our own reflections. A diamond is a tiny mirror: it helps us do just that.

Acknowledgments

Now we've reached the part of the book where it becomes especially difficult to avoid making diamond puns.

I have a gem of an agent in David Halpern, whom I am also very proud to call my friend, and who has given me the greatest gift I could ask for over the past ten years: the opportunity to write books for a living. I can't imagine working on a project without his support, ingenuity, enthusiasm, and very special way of saying "Calm down." I am excited for our kids to be best friends and love books together (but not too much). I thank the universe every day for delivering me into the hands of Jennifer Barth. Kind, cool, responsive, and encouraging, she is my dream editor, and getting a set of notes from her is like taking a master class in reading and editing. Without her, this book would be only half as good—but possibly twice as long. Thanks also to Erin Wicks, who provided insightful comments throughout this process and came up with a particularly clever solution to a problem with the last chapter. She also—willingly!—went into the trenches with me to create the book's photo insert. The fact that she still opens my e-mails is a testament to her patience. Once again, Kathy Robbins generously stepped in at exactly the right moment and helped me untangle some of the manuscript's trickiest knots. I treasure these experiences with her as the most fun an author can have while revising. My decade-long personal and professional relationship with Kathy is more evidence of my wonderfully good luck.

It's exhilarating—and a little bit nuts—to write about a century-old industry as an outsider, and I am beyond indebted to all of the archivists, authors, experts, historians, and jewelers who answered my e-mails, told me their stories, and passed along useful tidbits. I owe endless thanks to those who took time out of their busy lives to be interviewed: Jacob Arabo at Jacob & Co.; Ben Baller at IF & Co.; Lisa Bissell at Pure Grown Diamonds; Zandile Blay; Andrea Buccellati and Simona Meschi at Buccellati; Martin Chapman at the De Young Museum; Liz Chatelain at MVI Marketing; Elizabeth Doyle at Doyle & Doyle; Robyn Ellison at Rio Tinto Diamonds; Edward Jay Epstein; Ewing "Wing" Evans; Corinna Gilfillan at Global Witness; Robert Hazen at the Carnegie Institution for Science; David Johnson at De Beers; Jack Jolis; Douglas Kazanjian at Kazanjian Brothers; Greg Kwiat and Rebecca Selva at Fred Leighton; Joshua Lents at the Gemological Appraisal Laboratory of America; Karen Sampieri at Heritage Auctions; Peter Paul Scott; Sherman Shatz; Bevy Smith; Scott Sucher; John VerStandig; and Maurice "Mac" VerStandig. I am also so grateful to the people who answered my initial e-mail inquiries and often went on to coordinate these meetings: Shamin Abas and Natasha Berg at Shamin Abas Public Relations; Brandee Dallow and Sylvia Chapman at Rio Tinto Diamonds; Rena Gottleib at Tiffany & Co.; Clara Hatcher at the De Young Museum; James Her at IF & Co.; and Lindsey Ridell at Fred Leighton.

I think I regard archivists and librarians the way other people look at celebrities, and meeting Annamarie Sandecki at Tiffany & Co. was like spending time with the consummate rock star. Many thanks to

Amy McHugh and Christina Vignone at Tiffany & Co.; Erin Allsop at the Waldorf Astoria; Jeanette Berard at the Thousand Oaks Library; Lynn Eaton and Joshua Larkin Rowley at the Hartman Center for Sales, Marketing and Advertising History at Duke University Libraries; Michelle Frauenberger at the Franklin D. Roosevelt Presidential Library; Joe Hursey and Wendy Shay at the Smithsonian; Gene Morris at the National Archives; and Willie Thompson at the Scott and Zelda Fitzgerald Museum. Thank you to everyone at the Special Collections Research Center at Syracuse University Libraries, and the NYPL Manuscript and Archives Division. The New York Public Library Stephen A. Schwartzman Building remains one of my favorite places on earth.

I owe a particular debt to Kristin Mahan at the Gemological Institute of America, and to Augustus Pritchett at the GIA's Richard T. Liddicoat Gemological Library and Information Center, who both went above and beyond for me. I appreciate the people who helped in the often thankless job of coordinating photo permissions: Andy Bandit at 20th Century-Fox; Tad Bennicoff and Kealy Gordon at the Smithsonian; Pete Berenc at Getty Images; Tim Davis at Corbis Images; Rob Delap at the New-York Historical Society; Molly Rawls at the Forsyth County Public Library; Melissa Smith and Monica Whitehurst at Weber Shandwick; and Lauren Spoto at Pure Grown Diamonds. There is a special place in heaven for writers who help other writers, and I'd like to nominate Samantha Barbas, Zandile Blay, Robert Bates, Shelley Bennett, Edward Jay Epstein, Victoria Finlay, and Tom Zoellner for their wings. Friends—old and new—went out of their way to make introductions: Nicole Faux, Marc Foster, Sarah Gowrie, Kevin Krieser, Jove Oliver, Michael Palladino, and CK Swett. I'd also like to recognize Neal Dusedau and Louise Quayle for reading versions of the proposal and providing smart and helpful criticism, while also encouraging me to push forward.

My friends and colleagues at The Robbins Office are always so supportive of my alternate life as a writer, and two members of the extended family, both book experts in their own right—Richard Cohen and Lucinda Blumenfeld Halpern—are quick to brainstorm with me, offering keen advice and warm, genuine encouragement. And to the production team at HarperCollins: thank you for doing me the enormous favor of working on an accelerated schedule in order to accommodate my maternity leave.

I am blessed with a big and effusive family: my parents and sisters, Annette and Jay Bergstein, Pauline Bergstein and Jeff Wilson, Deanna Bergstein, and Allison Bergstein; my in-laws, Herb Rosenberg and Jean Rosenberg; my adopted brothers and sisters, of whom there are too many to name. You know who you are: scattered across the country, you are the people who make me think and laugh harder than anyone else does. I feel stupidly fortunate to know you. All of my parents have given me many, many gifts, but among the most important are the confidence and determination to keep doing what I love. You have set the example that hard works pays off, and that following my heart is never wrong. I can only hope to convey these same messages to my son.

Andrew Rosenberg gave me my first romantic diamond, and in doing so, made it very hard to be cynical about engagement rings, even as I discovered there was plenty to be cynical about. What can I say: every time I look down at my ring I think about the brilliant, solid, patient, and kindhearted man I married. Thank you so much for building a joyful life with me that makes room for this crazy job. And to our little guy who has kept me company for the past eight months: you are still a mystery and yet already so precious. I can't wait to know you and start this next chapter.

Notes

PREFACE

xv   slightly smaller than 16.5 mm: Blue Nile, "How to Determine
      Your Ring Size," http://www.bluenile.com/assets/chrome/pdf/
      ring_sizing_guide_o610-CA.pdf.

CHAPTER 1: THE BUYERS

The biographical details about Charles Lewis Tiffany come from the
      following sources:
      Clare Phillips, *Bejeweled by Tiffany: 1837–1987* (New Haven: Yale
          University Press, 2007).
      George Frederic Heydt, *Charles L. Tiffany and the House of
          Tiffany & Co.* (New York: Tiffany & Co., 1893). Available via
          Tiffany & Co. on Google Play, https://play.google.com/store/
          books/author?id=Tiffany%20and%20Company.
      "The Story of Charles Lewis Tiffany," CBS Radio, original air
          date March 4, 1937. Found on YouTube (accessed before
          12/13, link no longer available).
      "The Tiffanys: The Mark of Excellence," A&E *Biography*, DVD
          release date September 26, 2006.

Details about the Bradley-Martin ball come from the following
      sources:

Albin Pasteur Dearing, *The Elegant Inn* (Secaucus, NJ: L. Stuart, 1986).

James Remington McCarthy, *Peacock Alley: The Romance of the Waldorf-Astoria* (New York: Harper & Brothers, 1931).

Penny Proddow and Marion Fasel, *Diamonds: A Century of Spectacular Jewels* (New York: Harry M. Abrams, 1996).

"New York Fashions: Dress at the Bradley Martin Ball," *Harper's Bazaar*, February 20, 1897.

"The Bradley Martin Ball: A Wealth of Heirlooms, in Antique Jewels and Rare Old Laces, to Be Shown," *New York Times*, February 9, 1897, 3.

"Wore Crown Jewels: Mrs. Bradley-Martin Resplendent in a Bower of Beauty," *Boston Daily Globe*, February 11, 1897.

Records from the Waldorf Astoria archives, furnished by Erin Allsop, in-house archivist, by e-mail on 8/7/13.

6   where she declared $75,000 . . . in purchases: Shelley M. Bennett, *The Art of Wealth: The Huntingtons in the Gilded Age* (San Marino, CA: Huntington Library, 2013), 133.

7   they were able to advertise their selection of "French Jewelry": 1895 "Catalogue of Useful and Fancy Articles," Tiffany, Young & Ellis, 25. Found in the Tiffany & Co. archives in Parsippany, New Jersey, visited 5/14/2014–5/15/2014.

11   "A Corsage Bouquet—Two thousand six hundred and thirty-seven brilliants": "Crown Jewels of France: An Official Catalogue of the Collection," *New York Times*, April 24, 1887, 11.

11   "The sale of the Crown jewels seems to have excited more interest": "The French Crown Jewels: Opening of the Auction Sale in Paris," *New York Times*, May 13, 1887, 1.

12   "Of one thing we may be certain, that six months hence there will be ten times as many Crown jewels": Ibid.

12   With these photos in hand, some less scrupulous jewelers: Proddow and Fasel, *Diamonds*, 16.

13   "I know it would be an interesting piece of news": "The Crown

Jewel Sale Ended: More Purchases by New-Yorkers—The
Criers Dissatisfied," *New York Times*, May 24, 1887, 1.

16   hosts would request tiaras the way contemporary invitations
      denote black tie: Joan Younger Dickinson, *The Book of
      Diamonds: Their History and Romance from Ancient India to
      Modern Times* (New York: Avenel Books, 1965), 143.

16   The headpieces started at $150, or $4,123 today, in the Tiffany
      & Co. catalogue: From the 1895 Tiffany & Co. Blue Book, 215.
      Found in the Tiffany & Co. archives.

16   "There is no estimating the value of the rare old jewels to be
      worn": "The Bradley Martin Ball: A Wealth of Heirlooms," 3.

16   "'It is ridiculous to suppose'": Ibid.

16   Cornelia Bradley-Martin didn't pay $9,036.45 to the Waldorf:
      As enumerated on the bill in the Waldorf Astoria's archives,
      furnished by Erin Allsop, in-house archivist, by e-mail on 8/7/13.

CHAPTER 2: THE SEEKERS

23   Sensing opportunity, the shepherd demands just that: Stefan
      Kanfer, *The Last Empire: De Beers, Diamonds and the World* (New
      York: Farrar, Straus & Giroux, 1993), 26.

25   "Digging for diamonds never becomes dull drudgery":
      Gardner Fred Williams, *The Diamond Mines of South Africa*
      (London: Macmillan Company, 1902), 154 (page number
      refers to a digitized version available for B&N Nook).

25   they destroyed real diamonds by smashing them with
      hammers: Kanfer, *The Last Empire*, 29.

26   about 30,000 English pounds' worth of diamonds in 1870:
      Ibid., 30–31.

26   the widow charged two pounds a month: Ibid., 32.

27   news of the productive Colesburg Kopje echoed across the
      land: Williams, *The Diamond Mines*, 173–75.

28   he sold his farm to a group of investors for 6,000 guineas:
      Kanfer, *The Last Empire*, 34.

29   "Any day you may find a diamond that will astonish the world":
      Ibid., 60.

33   "Every acre added to our territory means in the future birth
      to some more of the English race": James Leasor, *Rhodes and
      Barnato: The Premier and the Prancer* (Smashwords, 2011), 99.

33   the ninety-eight separate claimholders still standing in
      Kimberley: Williams, *The Diamond Mines*, 278.

35   "When he talked Greek": Leasor, *Rhodes and Barnato*, 152.

38   "just children" and "emerging from barbarism": Ibid., 217.

39   a "gentle stimulus," Rhodes said: Ibid.

39   "When that great Kings return to clay": "Kipling's Tribute to
      Cecil Rhodes," *New York Times*, April 9, 1902, 1.

40   from 1898 to 1902, American imports of gemstones more
      than doubled: "Enormous Diamond Importations," *Keystone*
      (February 1903).

CHAPTER 3: THE LOVERS

Information on the history of diamond cutting comes from the
   following sources:
      Joan Younger Dickinson, *The Book of Diamonds: Their History
      and Romance from Ancient India to Modern Times* (New York:
      Avenel Books, 1965).
      Glenn Klein, *Faceting History: Cutting Diamonds and Colored
      Stones* (Bloomington, IN: Xlibris Corporation, 2005).
      Available on Google Books, https://books.google.com/
      books?id=AoZ2fb2-xZ8C.
      Diamonds: Myth, Magic, and Reality, ed. Jacques Legrand
      (New York: Crown Publishers, 1980).

Details about Sears history from the Sears Archives (http://www
   .searsarchives.com) as well as this profile: G. R. Clarke,
   "Dazzling and Sudden Success Not So Accidental as It Seems,"
   Chicago Daily Tribune, June 10, 1906, E3.

41   In 1902, 25,412,775.72 dollars' worth of diamonds and
      precious stones: "Enormous Diamond Importations."

42   "As the diamond does not appear to have been known to
      the ancients": George Frederick Kunz, *The Curious Lore of
      Precious Stones* (Philadelphia: J. B. Lippincott Co., 1913),
      321.

43   "[A] crusade is being started against the engagement ring":
      "Engagement Ring Trophies," *New York Times*, August 8,
      1894, 4.

44   "Are those five or six wedding rings all you have in stock?":
      "Trade Secret," *New York Times*, February 24, 1904, 8.

45   On October 11, Eleanor's twentieth birthday, the engagement
      became official: Hazel Rowley, *Franklin and Eleanor: An
      Extraordinary Marriage* (New York: Farrar, Straus & Giroux,
      2010), 37.

45   The ring's sizable center stone was cushion-shaped: Details
      about the ring supplied by Michelle Frauenberger at the
      Franklin D. Roosevelt Presidential Library and Museum, via
      e-mail on 7/23/13.

45   "the engagement ring is getting to be so indispensible": "Price
      of Engagement Ring Has Risen Forty Per Cent in the Last
      Three Years," *Chicago Daily Tribune*, November 27, 1904, F4.

47   the diamond was not in Persia at all but in Afghanistan: Ian
      Balfour, *Famous Diamonds* (Suffolk, UK: Antique Collectors'
      Club, 2009), 176.

48   Coeur recognized that the pretty, twentysomething Sorel:
      Dickinson, *The Book of Diamonds*, 52.

52   the price had doubled to a baseline of $20 ($540 today) for the
      least expensive version: 1894 Blue Book, 28, and 1908 Blue
      Book, 235, both available via Tiffany & Co. on Google Play.

54   "Without commenting on the foolishness of this statement":
      "Department Store Jewelry Advertising," *Keystone* 9 (October
      1897), NYPL bound edition, 822–23.

55   "The man or woman who is given an opportunity": Ibid.

CHAPTER 4: THE SHOW-OFFS

61   "A line of diamond fire in the square links of platinum":
      Evalyn Walsh McLean, *Queen of Diamonds: The Fabled Legacy of
      Evalyn Walsh McLean*, a commemorative edition of *Father Struck
      It Rich* (Franklin, TN: Hillsboro Press, 2000), 154.

63   Pierre had met and fallen for the American heiress Elma
      Rumsey: "Heiress to Wed Foreigner," *New York Times*,
      December 22, 1907, 1.

64   a 45.52-carat cushion-cut brilliant: Specifications of the Hope
      Diamond as per the GIA Colored Diamond Grading Report
      on the gem, posted on the Smithsonian's website, http://
      mineralsciences.si.edu/collections/hope/HopeGIAreport.pdf.

64   The price was $110,000: Richard Kurin, *Hope Diamond: The
      Legendary History of a Cursed Gem* (New York: HarperCollins/
      Smithsonian Books, 2006), 197.

69   There would be no deal. The spell was broken: Story and
      dialogue quotations from McLean, *Queen of Diamonds*, 167–73.
      Additional details sourced from Kurin, *Hope Diamond* and
      Victoria Finlay, *Jewels: A Secret History* (New York: Random
      House Trade Paperbacks, 2007).

69   he promptly reset it in a more contemporary corona: Kurin,
      Hope Diamond, 205–6.

70   The article also summarized the "sinister history": "J. R.
      M'Lean's Son Buys Hope Diamond," *New York Times*, January
      29, 1911, 1.

70   "Friends of the McLeans say to-night": "Sues the M'Leans for
      Hope Diamond," *New York Times*, March 9, 1911, 1.

71   "Should any fatality occur to the family": "Says M'Lean Drank
      Hope Diamond Toast," *New York Times*, March 10, 1911, 7.

71   "Although the writer has been connected with the jewelry
      trade press": T. Edgar Willson, letter to the editor, "The Hope
      Diamond: Editor Jewelers' Circular Writes of the Stories of
      Misfortunes," *New York Times*, February 9, 1911, 6.

72   McLean wore the Hope frequently: McLean, *Queen of Diamonds*, 174.

72   "the curse and the blessing fight it out together": Ibid., xviii.

74   gemstone historian Ian Balfour cites this as an argument: Balfour, *Famous Diamonds*, 131.

74   one fun theory places his uncle in the court of George IV: Ibid., 133.

76   Once again, the media took the opportunity to share details: "M'Leans to Keep the Hope Diamond," *New York Times*, February 2, 1912, 1.

76   Supposedly, Evalyn set the tone for the night: Finlay, *Jewels*, 329.

77   she drove to New York and hocked the Hope Diamond: McLean, *Queen of Diamonds*, 295.

78   "It is no use for anyone to chide me about loving jewels": Ibid., 290.

Additional details about the Hope Diamond furnished by a telephone interview with diamond cutter, historian, and investigator Scott Sucher on 10/30/13, and via his website: www .museumdiamonds.com.

CHAPTER 5: THE OPTIMISTS

80   "They'd wear diamonds on their ankles if it was stylish!": Lillian Ross, "The Big Stone II," *New Yorker*, May 15, 1954, 45.

80   "Why don't you marry me instead?": Ibid., 54.

81   On Forty-Seventh Street, jewelers had easier access to the trains: "Uptown Trade Movement: Diamond Merchants Desert Maiden Lane District for Midtown Section," *New York Times*, February 19, 1924, 31.

82   "The economic dictator of Germany increases the restraints": "Diamonds and Food," *New York Times*, March 23, 1916, 10.

83   while small diamonds, called melee, cost $450 to $500 . . . for a bundle of the same weight: "As to the Cost of Diamonds," *New York Times*, August 13, 1919, 17.

83   president Joseph Mazer predicted a solid future for diamonds: "Prohibition Helps Jeweler, Says Mazer," *Keystone* (October 1919), 131.

85   that Huntington had bought for a price of 100,000 francs: Details furnished from a visit to the Arabella Huntington Papers collection at Syracuse University, 9/23/13–9/25/13.

85   Her son and daughter-in-law were happy to sell almost the entire lot to the highest bidder: Bennett, *The Art of Wealth*, 271.

85   The investment required an enormous $1.5 million loan: Lillian Ross, "The Big Stone," *New Yorker*, May 8, 1954, 42.

86   calling the diamond business a "Cinderella world": Ibid., 45.

87   "Good baas [master], I have found it!": "Poor Prospector Finds 726-Carat Diamond Near the Site of the Cullinan Discovery in 1905," *New York Times*, January 18, 1934, 1.

87   Jonker's wife wrapped it in a cloth and slept with it hanging from her neck: "Jonker's Diamond Is Sold for £63,000," *New York Times*, January 19, 1934, 9.

88   Jonker promised that he would have continued employment: Ibid.

88   it reportedly cost the Diamond Corporation $15,000 a year: "$500,000 Bid for Diamond," *New York Times*, March 30, 1935, 8.

89   "It looked for all the world like a piece of the camphor ice": "Jonker Diamond Shown at Museum," *New York Times*, June 12, 1935, 23.

90   Years later, Kaplan admitted: Murray Schumach, "Lazare Kaplan, Diamond Dealer, Dies at 102," *New York Times*, February 14, 1986.

92   "It's perfect, Father!": "Jonker Diamond Is Cut in 3 Pieces," *New York Times*, April 29, 1936, 23.

92   Winston, ever the protective papa, confessed his wish to find one buyer: "Jonker Diamond Cut into 12 Perfect Gems; Set, Now for Sale, Valued at $2,000,000," *New York Times*, January 13, 1937.

92   with Kaplan griping that Winston took too much credit: Ross, "The Big Stone," 36.

93 The life of a workaday farmer didn't suit him: Balfour, *Famous Diamonds*, 152.

94 its 1932 issue featured "Gifts one dollar upward": Judy Rudoe, *Cartier: 1900–1939* (New York: Harry N. Abrams, 1997), 38.

95 "Jewelry is something more than adornment": New York Public Library World's Fair archives box 2129.

95 As visitors entered in groups: Description from NYPL World's Fair archives box 443.

96 Tiffany & Co. staged a particularly impressive display: "Articles of Jewelry to Be Exhibited by Tiffany & Co. at Their Exhibit in the House of Jewels, New York World's Fair," 1939, New York World's Fair Scrapbook; found in the Tiffany & Co. Archives.

96 The 1939 World's Fair was a popular success: "1939 Fair Closes; Seen by 26,000,000; Plans Laid for '40," *New York Times*, November 1, 1939, 1.

Details about hours of the House of Jewels from NYPL World's Fair archives.

Other details about Lazare Kaplan and the Jonker from the Lazare Diamonds website, www.lazarediamonds.com/AboutUs/OurHistory.

CHAPTER 6: THE SELLERS

The story of "A Diamond Is Forever" is put together from recollections found at the Duke University Libraries (J. Walter Thompson collection) and from the Smithsonian Institution Archives (N. W. Ayer Advertising Agency Records, 1849–1851, 1896–1996). Particularly useful was an interview conducted with Frances Gerety by one Howard Davis, who was working on putting together an oral history of N. W. Ayer & Son. It was never published.

102 "the most important event of [his] Ayer career": As quoted in an unpublished interview with Paul Darrow. Found in the Smithsonian's N. W. Ayer & Son archives, visited 2/27–2/28/13.

103 they came up with four objectives for the ads: "Art and Ayer: A Tribute to Excellence," *Diamond Line*, Christmas issue 1986. Found in the Smithsonian's Ayer & Son archives.

103 research showing that men typically shopped for engagement rings alone: According to an N. W. Ayer & Son press release, "Mr. Cupid Now Buying the Diamond 'Solo.' " Found in the Duke University Libraries' J. Walter Thompson (JWT) archives, visited 2/21–2/23/13.

103 In order to save money, the art department had settled on a two-color palette: "Art and Ayer: A Tribute to Excellence."

104 less than $80 for the jewel that would ultimately beget his future: Edward Jay Epstein, *The Rise and Fall of Diamonds: The Shattering of a Brilliant Illusion*, ebook edition (New York: Simon & Schuster, 1982; Smashwords, 2011), 147. Citations refer to the 2011 ebook edition.

107 diamond sales rose 55 percent: Ibid., 76.

107 "Girls have stopped waiting for their fiancés": "People and Ideas: Accelerated Wedding Plans," *Vogue* 101 (1943), 62.

107 "First comes the engagement ring of gold": "Hound, Pennywise," *Vogue* 99, no. 7 (April 1, 1942), 33.

108 "A wedding ring manufacturer announces": Found in the Duke JWT archives.

108 "The earth may heave. Stars may fall": "Diamonds Remain Symbol of Man's Ideal of Romance," *Atlanta Constitution*, March 9, 1940, 14.

108 "The solitaire, or one fine diamond in a plain setting, is back in fashion": "Solitaire Returns to Favor for Engagement Rings," *Washington Post*, February 15, 1941, 13.

109 "Although wearing a ring on the left hand does not announce an engagement": Emily Post, "Solitaire Would Suggest an Engagement," *Daily Boston Globe*, October 27, 1940, B51.

109 "It is the right of the bride to choose what she likes best!": Emily Post, "Are Conventions Superficial," *Daily Boston Globe*, September 12, 1940, 24.

109   remarked that, unfortunately, Emily Post couldn't be bought:
        Memo from George D. Skinner to Dorothy Dignam, "Etiquette
        Brochure on Engagement Ring," dated August 14, 1959. Found
        in the Duke JWT archives.

110   "You will find more than starry gems and ornaments": De
        Beers ad, *Vogue* 98, no. 8 (1941).

110   "Time and circumstances and thousands of miles": De Beers
        ad, *Vogue* 99, no. 8 (1942).

115   "I shudder to think of what might have happened if a great
        line had been demanded": Letter from Frances Gerety to
        Mr. F. Bradley Lynch, dated October 20, 1987. Found in the
        Smithsonian's Ayer & Son archives.

115   the painter was feeling sick after trying an experimental
        diet that involved eating human hair: From an interview
        with Charles Coiner in his personnel file. Found in the
        Smithsonian's Ayer & Son archives.

CHAPTER 7: THE QUEENS

119   "I've made up my mind to meet him": Michael Bloch, ed.,
        Wallis and Edward: Letters 1931–1937 (New York: Simon &
        Schuster, 1992), 33.

119   "Something ought to be done about the lights": Ibid., 46.

120   she asked Simpson to "look after" the prince for her: Ibid., 89.

122   "Capetown's interpretation of the color bar is the most liberal":
        G. H. Archambault, "South Africa Roars a Welcome to the Royal
        Family of Britain," *New York Times*, February 18, 1947, 1, 22.

122   When increasingly critical reports started surfacing:
        "Elizabeth Will Not Get Gift of 400 Diamonds—Only 21," *Los
        Angeles Times*, January 30, 1947.

123   "What a wonderful present, I am delighted!": "South Africa Rains
        Diamonds on Royal Family," *Chicago Tribune*, April 19, 1947, 1.

125   "Enchanting as a fairy tale, [Princess Elizabeth's] wedding
        had the side effect": Proddow and Fasel, *Diamonds*, 97.

125   her collection of high-end baubles soon became the talk of the London social circuit: Anne Sebba, *That Woman: The Life of Wallis Simpson, Duchess of Windsor* (New York: St. Martin's Griffin, 2013), 112.

126   "I certainly was no beauty": Bloch, *Wallis and Edward*, 102.

127   "True we are poor and unable to do the attractive amusing things in life": Ibid., 179.

127   Some speculate that King Edward approached Ernest: Ibid., 154–56. This refers to the account of Ernest Simpson's friend Bernard Rickatson-Hatt, who said he witnessed a confrontation between the two men.

128   the actual crowns that he and Simpson would no longer have at their fingertips: "Edward Gives Up Crowns Noted in British History," *Chicago Tribune*, December 11, 1936, 9.

128   "bitter question which secretly agitated high British circles": "Other Women Envy Wally's Jewels," *Daily Boston Globe*, December 20, 1936, C3.

128   "What is it that puts the Duchess of Windsor at the top of every 'best dressed' list": "Duchess of Windsor Sets Her Own Style Barometer," *Daily Boston Globe*, April 25, 1948, A2.

128   Another Van Cleef piece arrived on Simpson's fortieth birthday: From the Van Cleef & Arpels website, http://www.vancleefarpels.com/ww/en/la-maison/icons/Iconic-clients/the-duchess-of-windsor.html.

129   which he had purchased from industrialist Morton F. Plant for the price of $100: Martin Chapman, *Cartier and America* (New York: Prestel Publishing, 2009), 23.

129   There, he worked alongside artistic director Jeanne Toussaint: Jane Barry, "Parisian Designs Unusual Settings for Jewels," *Christian Science Monitor*, May 17, 1951, 14.

129   Toussaint, who during the Nazi occupation integrated subtle patriotic imagery into her designs: Chapman, *Cartier and America*, 100.

129 Anne Sebba contends that the two women . . . understood each other: *That Woman*, 270.

130 This one—an articulated bracelet with the classic diamond-and-onyx skin: Details about panther skin and the panthers from Rudoe, *Cartier: 1900–1939*, and Chapman, *Cartier and America*.

130 As early as 1946, when her English country apartment was burglarized: "Part of Wally's Stolen Jewels Found on Links," *Chicago Daily Tribune*, October 18, 1946, 3.

130 In 2010, when Sotheby's London hosted a posthumous sale of the duchess's jewelry: Tamara Cohen, "Is Madonna the New Owner? Mrs. Simpson's Panther Bracelet Sells for a Record Breaking £4.5m," *Daily Mail*, December 1, 2010, http://www.dailymail.co.uk/femail/article-1334499/Is-Madonna-new-owner-Wallis-Simpsons-4-5m-panther-bracelet.html.

131 "one of the most unusual enterprises in the glittering half-mile": "Fifth Avenue: A Jeweler's Showcase," *Washington Post*, March 6, 1953, 48.

131 At one point, he lobbied the White House to put together a national collection of jewels: Kurin, *Hope Diamond*, 247.

132 There were plenty of potential buyers, but Harry Winston swept in and offered to take the whole lot: Balfour, *Famous Diamonds*, 141.

132 Soon after he acquired it, Winston and his wife, Edna, decided to fly home: Ibid., 142.

133 "To most people's amazement, the thing turned out to be dark blue instead of white": "Hope Jewel Displayed in Public Cause: Famed Blue Diamond Shines for New York United Hospital Fund," *Hartford Courant*, November 25, 1949, 15.

CHAPTER 8: THE STARS

138 "If I rattle with clips from Flato": From Cole Porter's "Yes Yes Yes," in *The Complete Lyrics of Cole Porter*, ed. Robert Kimball (New York: Da Capo Press, 1983), via Google Books.

138 Ettinger got her start at MGM, and had rebranded an aspiring film-processing client as Technicolor: Sheila Weller, *Dancing at Ciro's: A Family's Love, Loss, and Scandal on the Sunset Strip* (New York: St. Martin's Press, 2003), p. 98.

139 "wore a wedding ring and a diamond engagement ring": Louella Parsons, "Screen Snapshots," *Washington Post*, February 4, 1940.

139 "the engagement ring Jeffrey Lynn gave Dana Dale is a star sapphire set with diamonds": Louella O. Parsons, "Close Ups and Long-Shots of the Motion Picture Screen," *Washington Post*, October 10, 1940.

140 Clearly, someone who signed Maggie Ettinger's paychecks thought so: Janine Roberts, *Glitter and Greed: The Secret World of the Diamond Cartel* (New York: Disinformation Company, 2003), 350.

140 "Mickey [Hargitay], my husband, gave me that diamond with his love": "The Day I Got My Diamond," *Motion Picture*, April 1959, 68.

141 On her twenty-fifth anniversary at Ayer, her colleagues presented her with the perfect present: From a letter found in the Smithsonian's Ayer & Son archives.

142 love at "second sight," she quipped winningly: "Grace Glows: 'Love Was Never Like This,' " *Newsday*, January 9, 1956, 4.

143 Within the year, he'd named the house the "Official Supplier of the Principality of Monaco": From the Van Cleef & Arpels website, http://www.vancleefarpels.com/ww/en/la-maison/icons/Iconic-clients/hsh-princess-grace-of-monaco.html.

144 "a bon voyage basket that would be a credit to Tiffany's window": "Grace Kelly's Last Fling in Movie High Society," *New York Herald Tribune*, April 8, 1956.

145 Columbia Pictures distributed the twenty-seven-minute color film across 3,500 theaters that year: Vicki Howard, *Brides, Inc.: American Weddings and the Business of Tradition* (Philadelphia: University of Pennsylvania Press, 2008), 58.

145 the film opened with Mary scribbling in her diary, wondering how to describe "the happiest day of [her] life": Loland Baxter,

"A Diamond Is Forever," shooting script, September 1952. Found in the Duke JWT archives, sent by Joshua Larkin Rowley via e-mail on 6/12/14.

146 Hannaford hadn't always dreamed of being a "career woman": Edyth Thornton McLeod, "Poise and Personality Aid in Career After Forty," *Brooklyn Eagle*, September 19, 1951, 17.

147 The very first question was a doozy: "Is a diamond a good investment?": "Is Investment in Diamonds Worthwhile?" *Chubbs Reporter*, July 16, 1952, 1.

148 When asked, at a session of the Fifty-Second Annual New York State Retail Jewelers' Convention in Binghamton: Marilyn Young, " 'Diamond Girl's Best Friend' Is Opinion of Gem Lecturer," *Binghamton* (NY) *Press*, May 8, 1961.

148 "The most important diamonds are those worn on the third finger, left hand of American girls": "Hannaford Speaks on Famous Gems," *Benson High News*, October 21, 1949, 1, http://bhs.stparchive.com/Archive/BHS/BHS10211949p01.php.

149 "When we look at the characteristics of the women who have not received a diamond engagement ring": From a report, "The Gem Diamond Consumer Research." Found in the Duke JWT archives.

149 "Those born during the post-war baby boom will be reaching the age for high school graduation": Stewart Kampel, "Jewelers Expect Improvement to Pave Way for Glittering '63," *New York Times*, January 8, 1962, 86.

150 The real Tiffany & Co. welcomed its role on the big screen: Aubry D'Arminio, "Audrey Hepburn Accessorizes with the 128.54 Carat Tiffany Diamond in 1960," *JCK*, April 2013, http://www.jckonline.com/2013/03/25/audrey-hepburn-accessorizes-12854-carat-tiffany-diamond-in-1960.

150 In appreciation, the studio hosted a private screening for employees: Memo from W. F. Stanton dated July 19, 1961, from Tiffany & Co. notices, volume 6, 1959–1961. Found in the Tiffany & Co. archives.

151   In November 1956—just one month after one of
Schlumberger's pins appeared on the cover model: "Fashion:
The Tiffany Diamond and Its Wardrobe of Settings," *Vogue*,
November 15, 1956, 92–93.

151   The Schlumberger necklace looked perfect on slim, elegant
Hepburn, who tried it on for photographers: D'Arminio,
"Audrey Hepburn Accessorizes."

152   "It might have been more attractive to have a young woman
wear it": Nan Robertson, "Diamond Worth $583,000 Is
'Comfortable' to Wear," *New York Times*, July 17, 1957, 43.

152   "We won't sell diamond rings for men because we don't like
it": Lisa Hammel, "10 Years as Tiffany's Alarm Clock," *New York
Times*, September 10, 1965, 43.

153   "Our younger customers are becoming educated to good
taste": David Kahn, "Wider Price Range Broadens Tiffany's
Net," *Newsday*, July 5, 1960, 16C.

153   The new ads were entirely different: Advertising Collection,
1963 and 1965. Found in the Tiffany & Co. archives.

154   Once again, as he did with the Jonker, he had the diamond
sent by US mail: Milton Bracker, "The Hope Diamond Is Off in
the Mail," *New York Times*, November 9, 1958, 56.

CHAPTER 9: THE WINNERS

Information about Elizabeth Taylor's life from "Elizabeth Taylor:
Facets," A&E *Biography*, original air date March 16, 2003.

162   In 1949, in its "Italian Handbook," *Vogue* referred to Bulgari:
"Italian Handbook," *Vogue*, May 15, 1949, 69.

163   When he opened at the new, bigger location, he labeled
S. Bulgari an "Old Curiosity Shop": Daniela Mascetti and
Amanda Triossi, *Bulgari* (New York: Abbeville Press, 1996), 15.

164   earned himself the title in Italy of "Jeweler to Kings": Ibid., 22.

165   "Bulgari's nice little shop": Elizabeth Taylor, *My Love Affair
with Jewelry* (New York: Simon & Schuster, 2002), 56.

167   At the store, Gianni revealed "green flames": Ibid., 58–59.

167   "It's really like getting two pieces in one": Ibid., 59.

167   at 23.44 carats: From the Bulgari website, http://www.bulgari
        .com/en-us/heritage_starring_pieces_discover.

169   In early 1944 Winston received an invitation: Harry Winston,
        Harry Winston (New York: Rizzoli, 2012), foreword by André
        Leon Talley, 13.

171   as he told the *Boston Globe* soon after *Vogue* proclaimed him
        an eligible bachelor: Marian Christy, "Bulgari: Jewel of a
        Bachelor," *Boston Globe*, August 14, 1969, 30.

171   "I'm just tickled to death [that Bulgari is in New York]": Judy
        Klemesrud, "Bulgari Opens Jewelry Boutique with Lunch, Tea
        or Dinner," *New York Times*, December 14, 1971, 58.

172   "We like to think of our jewelry as art": Ibid.

CHAPTER 10: THE MEDDLERS

173   Taylor wore the diamond on her ring finger and enjoyed the
        irony of owning it as a proud Jewish woman: Taylor, *My Love
        Affair with Jewelry*, 49.

175   "one tiny slip can mean a million-dollar loss": This and other
        details from an article by Murray Schumach, "Tools and Skill
        Cleave a Diamond," *New York Times*, December 21, 1966, 41.
        Additional details about the Taylor-Burton Diamond come
        from Balfour, *Famous Diamonds*, and Taylor, *My Love Affair with
        Jewelry*.

175   Ames bought the stone for $500,000 (almost $3.7 million
        today): Kenneth Schwartz and F. Peter Model, "Diamonds Are
        a Girl's Best Friend," *Washingtonian*, January 1980, 72–77. I
        was pointed toward this article by a reference in Epstein, *The
        Rise and Fall of Diamonds*.

175     $30,000 a year ($212,500 today), according to some sources:
        Epstein, *The Rise and Fall of Diamonds*, 147.

176   The catalogue was called "Precious Stone Jewels": Charlotte

Curtis, "Cartier Pays a Record $1,050,000 for Somebody's Diamond," *New York Times*, October 24, 1969, 50.

176   "I guess it's for the man who has everything": Ibid.

177   "he was sure the company would have no difficulty selling it": Joseph Zullo, "69.42 Carat Diamond Sold for $1,050,000," *Chicago Tribune*, October 24, 1969, 1.

177   "not a free agent": Curtis, "Cartier Pays a Record $1,050,000," 50.

177   leading some journalists, like investigative reporter Edward Jay Epstein, who wrote about diamonds in the early 1980s: Epstein, *The Rise and Fall of Diamonds*, 148.

177   Taylor later revealed that they paid an additional $50,000 above the high bid: Taylor, *My Love Affair with Jewelry*, 93.

178   "Probably there were lots of people who have never been in here before": "Thousands Jump at Chance to See Expensive Gem," *Hartford Courant*, October 26, 1969, 4B1.

178   the syndicated gossip column "Suzy Says" outed Harriet Annenberg Ames as the diamond's original owner: "Suzy Says," *Chicago Tribune*, October 30, 1969, B1.

178   "The peasants have been lining up outside Cartier's this week": "The Million-Dollar Diamond," *New York Times*, November 1, 1969, 32.

178   "Despite the fact that the Burtons have earned their money": "Gee, It's Be-Yoo-Tiful!" *Hartford Courant*, October 28, 1969, 26.

179   One woman even noted that diamonds didn't go well with the newest styles: Curtis, "Cartier Pays a Record $1,050,000," 50.

179   the younger generations were having trouble scraping together the money for engagement rings, and were forgoing diamonds: "Pearly Ring Seals Romance," *Hartford Courant*, June 11, 1964.

179   "Bored with Diamonds? Consider Jade": Patricia Peterson, *New York Times*, February 6, 1972.

179   "Nowadays, a Girl's Best Friend May Indeed Be Jade": Angela Taylor, *New York Times*, November 22, 1971, 42.

180   "Mother's life seems just as fantastic to me as it must seem to
       everyone else": "Diamonds Are for Never: Liz's Son Takes to
       the Hills," *Los Angeles Times*, May 8, 1972, 2.

181   Two more African mines followed, both of which posed
       significant engineering challenges: Details about Namibian
       and Lesotho mines from Epstein, *The Rise and Fall of
       Diamonds*.

183   They innovated the craft by implementing an assembly-line
       system: "Diamonds—An Export Success," *Jerusalem Post*,
       October 18, 1966, vii.

184   To his friends, Monty Charles was a convivial chap: "Monty
       Charles," obituary, *Telegraph*, September 6, 2004, http://www
       .telegraph.co.uk/news/obituaries/1471059/Monty-Charles
       .html.

186   "He said they got into this business": Memo from Don
       Thompson to Clem Kressler, dated October 18, 1962. Found in
       the Duke JWT archives.

186   "I think calling small diamonds 'miniature diamonds' makes
       about as much sense": Memo from Kathie Horan to George
       Skinner, dated June 12, 1962. Found in the Duke JWT archives.

187   "Only 14% of married women have received a new diamond
       gift of any kind since their marriages": N. W. Ayer memo, "A
       Program to Increase the Consumption of Small Melee in the
       U.S. Market," undated. Found in the Duke JWT archives.

187   Perhaps jewelers might create special pieces to commemorate
       each milestone: Memo from Mildred Kosick to Donald C.
       Thompson, "Comments on Melee Report from Plans Dept,"
       dated January 16, 1963. Found in the Duke JWT archives.

188   now surveys showed a notable change, not just in buying
       habits but also in the attitudes of the present wedding
       generation: Report for N. W. Ayer, "The Market for Diamond
       Jewelry, United States," dated 1982. Survey credited to Karl
       H. Tietjen Research Associates. Found in Duke University's
       Jeffrey Collection of Diamond Promotional Materials.

188   De Beers admitted in the *Wall Street Journal* that in 1975 diamond sales had declined by 6.5 percent: "De Beers Diamond Sales Declined 6.5% Last Year," *Wall Street Journal*, January 13, 1976, 33.

189   The British were already coming around, with over 60 percent of grooms-to-be in 1966 feeling pressure to go to the jewelry store: "De Beers Aims Campaign at Japanese Brides," *Chicago Tribune*, December 30, 1966, B8.

189   "European jewelers are starting a campaign to make the diamond a more significant part of betrothal and marriage as it is in this country": Memo from Dorothy Dignam, dated May 22, 1961. Found in the Duke JWT archives.

191   Research showed that before the mid-1960s when De Beers kicked off the campaign, only 6 percent of Japanese brides received a diamond engagement ring: From a 1982 presentation to the US Carat Club. Found in the Duke JWT archives.

CHAPTER 11: THE INVENTORS

Details about the Lemoine case from various articles at the time; I was tipped off to it by Robert M. Hazen, *The Diamond Makers: A Compelling Drama of Scientific Discovery* (Cambridge: Cambridge University Press, 1999), and Kanfer, *The Last Empire*.

198   "Fraud will be impossible": "Lemoine Prepares to Make Diamonds," *New York Times*, May 31, 1908, 16.

198   " 'L'affaire des diamants' . . . [made] the famous Dreyfus case seem flat": "Is Lemoine the King or Knave of Diamonds?" *New York Times*, February 2, 1908.

199   "I have often encountered the belief that the successful solver": As quoted in Hazen, *The Diamond Makers*, 56.

200   In 1950, it put together a team and launched the top-secret Project Superpressure: Ibid., 97.

201   With the help of his colleagues in the workshop, he'd built his machine in the off-hours: Ibid., 118.

201   "So far, [the GE lab] has made only $10 worth of diamonds":
        "Giant General Electric Press Turning Out Real Diamonds,"
        Baltimore Sun, February 16, 1955, 3.

201   shares of General Electric on the New York Stock Exchange
        showed unusually high activity: William L. Laurence,
        "Diamond Is Created by G. E. Scientists," *New York Times*,
        February 16, 1955, 32.

204   "It was one of the real ones. So sorry": Telephone interview
        with John VerStandig (JVS), Helen Ver Standig's son,
        3/4/15.

206   "Now understand, this would be like Kenny Feld": JVS to RB,
        2/24/15.

208   One 1975 ad that ran in the *New York Times* was written in
        Madame Wellington's voice: Wellington Jewels ad ("There
        Ain't No Depression"), *New York Times*, January 12, 1975, 37.

208   "Honey, I'm responsible for more cheap weekends than
        any madam in the country": Stephanie Mansfield, "Ersatz
        Diamonds Are This Girl's Best Friend," *Washington Post*,
        December 12, 1981, 16.

209   it was an over-$20-million-a-year enterprise: JVS to RB,
        4/19/15.

210   "My customers love it when they are being robbed": Lee
        Wohlfert-Wihlborg, "Diamonds (Fake Ones) Are This Girl's
        Best Friend—Just Call Her Madame Wellington," *People*,
        November 23, 1981, http://www.people.com/people/archive/
        article/0,,20080751,00.html.

211   with De Beers ordered to pay GE an undisclosed sum: Epstein,
        The Rise and Fall of Diamonds, 101.

211   rumored to be somewhere between $8 million and $25
        million, plus royalties: Hazen, *The Diamond Makers*, 180.

211   "[didn't] know whether it will ever become possible, in the
        future": Gene Smith, "Laboratory-Produced Diamond of
        Gem Quality Made by G.E.," *New York Times*, May 29, 1970, 36.

211   "We . . . congratulate them on that achievement": Reuters, "De

Beers Sees No Threat in G.E. Diamond Discovery," *New York Times*, May 30, 1970, 23.

CHAPTER 12: THE ILLUSIONISTS

215   "If you own a Picasso, he's dead": William H. Jones, "A Free Spirit, a Counterfeiter: Madame Wellington's Gift for Self-Promotion," *Washington Post*, April 30, 1978, F1.

215   "Look at it this way: when the jeweler sells you a stone, he's making a profit": Lori Grey, "Is a Real Diamond Also a Girl's Best Investment?" *Baltimore Sun*, December 5, 1982, 78.

217   "I realized there was a whole cartel controlling the world of diamonds": In-person interview with Edward Jay Epstein in New York City, 7/10/14.

217   "What [De Beers] succeeded in doing, which is one of the most brilliant things I know of in any industry": Ibid.

219   "What they said to me," Epstein remembers, "is more or less this": Ibid.

219   "I remember a good friend of mine . . . he said, your diamond book changed my life": Ibid.

222   "This week Lady Diana Spencer renounced for life her privacy, independence and freedom": Jane Ellison, "Lady Diana Spencer: From the Silver Spoon to the Royal Crown," *Los Angeles Times*, March 1, 1981, F1.

222   a Mr. Stephen Barry appeared on the American television news program *20/20* sharing gossip from Buckingham Palace: Alex Brummer, "Why Diana's Ring 'Made the Queen's Eyes Pop,'" *Guardian*, March 18, 1983, 6.

222   To that end, De Beers announced within industry circles that it would be increasing its advertising budget: Karen Berman, "De Beers to Spend $20M in 1981: Small Stones Get Push," *National Jeweler*, February 1, 1981, 1.

223   "attention will be focused on the diamond": "De Beers Aims to

Upgrade Engagement Ring Caratage," *National Jeweler*, March 16, 1981, 81.

223 "She can't flaunt a fur on the Côte d'Azur": De Beers ad, *Vogue* 172, no. 6 (1982).

224 "Some of your customers may have little or no idea of what they can expect to spend": "The Diamond Engagement Ring," video, Diamond Promotion Service, 1980s (exact date unknown). Found in Duke University's Jeffrey Collection of Diamond Promotional Materials, box 2.

225 "the diamond is a symbol of their love": Ibid.

227 Even Marylou Whitney (of *the* Whitneys) admitted to doing it: Joan Kron, "If Diamonds Are Girl's Best Friend, CZs Are Good Pals," *Wall Street Journal*, June 15, 1984, 1.

227 "We make the material as though it were fertilizer": United Press International, "Cubic Zirconia Booming: Look-Alike Sells at Only Fraction of Diamond Cost," *Los Angeles Times*, April 1, 1986, OC_C12.

228 "We place the emphasis on fun": "Diamond-Duplicates Hard to Distinguish," *Los Angeles Times*, May 10, 1979, WS12.

228 "Debating" confessed that his girlfriend wanted a "big rock," but he couldn't afford it: Abigail Van Buren, "Don't Let Gems Put Engagement on the Rocks," *Chicago Tribune*, August 30, 1983, B4.

CHAPTER 13: THE MASTERS

230 Brady grew up poor in Lower Manhattan: "Times Topics: Diamond Jim Brady," *New York Times*, http://topics.nytimes .com/top/reference/timestopics/people/b/diamond_jim_ brady/index.html.

231 "It is a known and recognized fact that many wealthy jewelers and diamond dealers make little display of jewelry: " 'Sparkling Billy' Craig," *Jeweler's Circular* 83, no 1 (October 19, 1921), 99.

231 the ideal white diamond had been described colloquially:

Telephone interview with Robyn Ellison (RE), communications manager for Rio Tinto Diamonds, 10/23/14.

232   "You can consult your jeweler to have them custom-made": "Brown Diamonds a Keynote in Men's Fashions," *National Jeweler*, October 1941, 144–45.

233   Ayer conducted market research and found that 75 percent of men's jewelry purchases were actually made by women as gifts: Jeffrey J. Csatari, "De Beers, Intergold Expand Men's Jewelry Promos," *National Jeweler*, April 16, 1984, 79–81.

234   In 1972, a group of five mining companies contributed 20,000 Australian dollars each: Rio Tinto Mining Company of Australia, *Barramundi Dreaming: The Argyle Diamonds Story* (2012), 17.

234     $450 million, when all was said and done: From the Argyle mine website, http://www.argylediamonds.com .au/argyle_history.html (accessed before 11/14, link no longer valid).

236   "I'm perhaps the only person in our organization who can remember times": Robert L. Muller, "De Beers Says It Can Survive Without Having to Cut Diamond Prices," *Wall Street Journal*, May 11, 1982, 35.

236   In 1986, Argyle was expected to produce twenty-five million carats: Mark A. Klionsky, "Argyle Mine to Expand World Diamond Output 50%," *National Jeweler*, April 16, 1984, 49, 56.

236   A spokesperson for De Beers, in conversation with *National Jeweler*: Ibid.

237   And a great number were unmistakably brown: Telephone interview with Liz Chatelain (LC), president of MVI Marketing, 10/6/14.

237   "It was a very fractious relationship [with De Beers]": RE to RB, 10/23/14.

237   However, Belgian customs wasn't entirely prepared for the sheer quantity: Ibid.

238   it hired a Boston market research and consulting firm: LC to RB, 10/6/14.

238   "My partner and I looked at each other and said, wow, this is
        going to be difficult": Ibid.

239   While some jewelers remained skeptical even after seeing
        MVI's traveling collection: Ibid.

240   In 1991, champagne diamonds were about 30 percent cheaper:
        Michael Richardson, "Australian Miners Peddle a Browner
        Shade of Sparkle," *New York Times*, December 23, 1991.

241   In 1990, the Agra Diamond, cushion-cut and weighing in at
        32.24 carats: Christie's Hong Kong, "The Perfect Pink: An
        Extremely Rare 14.23 Carat Pink Diamond Leads Christie's
        Jewels Sale in Hong Kong on 29 November," news release,
        September 16, 2010, http://www.christies.com/presscenter/
        pdf/2010/JLS_HK_291110.pdf.

242   a promotion originally hatched by Arden and Hollywood
        jeweler Paul Flato: Elizabeth Irvine Bray, *Paul Flato: Jeweler to
        the Stars* (Suffolk, UK: Antique Collectors' Club, 2010), 40.

242   making up one-tenth of 1 percent of Argyle's annual
        production: Rio Tinto, *Barramundi Dreaming*, 121.

242   "a fabulous fluke of nature": RE to RB, 10/23/14.

242   Seven years later, in 1991, when the Australians renewed their
        contract with De Beers: Ibid.

244   the average price dropped to $300: LC to RB, 10/6/14.

245   On the contrary, when the opportunity came up in 2000 for De
        Beers to make a takeover bid: "De Beers Gives Up on Ashton,"
        New York Times, November 15, 2000.

CHAPTER 14: THE STUNNERS

Biographical information about Jacob Arabo from an in-person
        interview with him at the Jacob & Co. flagship in New York
        City on 4/6/15.

251   "It should never be a surprise that a lot of the status symbols
        that equate with success": Telephone interview with Bevy
        Smith (BV), 8/11/14.

252 DJ, brand manager, and hip-hop entrepreneur Peter Paul Scott credits influences: Telephone interview with Peter Paul Scott (PPS), 8/11/14.

253 "With the hip-hop community, we remix the way things are used": BS to RB, 8/11/14.

253 The gamble turned out to be a very good one: Chuck Philips, "The Uncivil War: The Battle Between Establishment and Supporters of Rap Music Reopens Old Wounds of Race and Class," *Los Angeles Times*, July 19, 1992.

255 an item that Los Angeles jeweler Ben Baller estimates cost about $28,000: Ben Baller, "The 50 Greatest Chains in Hip Hop," *Complex*, December 5, 2011.

255 "People had the most magnificent cartoon-character symbols": BS to RB, 8/11/14.

257 "That's how big [Jacob] got": PPS to RB, 8/11/14.

259 "I want the bling-bling": Mike Wise, "It's Time to Chase the Moment in the Sun," *New York Times*, April 22, 2000.

259 "So many blinding diamonds and the words 'Bling! Bling!' were inscribed on the side": Thomas George, "Rice Hoping to Come Full Circle," *New York Times*, November 10, 2000.

260 In 2002, CNN anchors were encouraged to use slang like "bling bling": Jim Rutenberg, "A New Vocabulary at Headline News," *New York Times*, October 7, 2002.

262 for a whopping $1.2 million: In-person interview with Ben Baller in Los Angeles, 9/24/14.

263 "I was like a dude out of jail": Ibid.

263 "Give me all your ideas": Ibid.

264 "I deliver and people's eyes are like, oh wow, man": Ibid.

264 "We bring Cartier to them": Ibid.

CHAPTER 15: THE CRITICS

268 Then, in an effort to finance its ongoing rebel campaign, UNITA seized control: Global Witness, *A Rough Trade: The Role*

of Companies and Governments in the Angola Conflict (London: December 1, 1998), 3.

269   "To the millions of Angolans who have survived the repeated years of conflict": Ibid., 4.

270   "the experts who work on behalf of the Ministry of Economic Affairs are . . . generalists": Ibid., 10.

271   Rapaport, the son of Holocaust survivors, had an instinctive, emotional response: For a more comprehensive account of Rapaport's role in the conflict diamond issue, see Alicia Oltuski, *Precious Objects: A Story of Diamonds, Family, and a Way of Life* (New York: Scribner, 2011).

272   RUF also had an extremely grisly signature: Greg Campbell, *Blood Diamonds: Tracing the Deadly Path of the World's Most Precious Stones* (Boulder, CO: Westview Press, 2002), 15.

274   "I think a lot of the industry and the governments thought of the fur trade": Telephone interview with Corinna Gilfillan (CG), 7/11/14.

275   "any trader that has dealt with these illicit diamonds will be banned": "Diamond Industry Acts to Halt Trade in Illicit Gems from Africa," *New York Times*, July 20, 2000.

275   At the height of the crisis, it was estimated that between 4 percent and 15 percent of diamonds: Global Witness, *The Truth About Diamonds* (London: November 15, 2006), 2.

276   "I think one of the real challenges of the scheme": CG to RB, 7/11/14.

276   that assumes other participating governments not only are robustly implementing their own regulations: Ibid.

277   "Since the grisly execution of Operation Clean Sweep in 1996": Campbell, *Blood Diamonds*, xviii.

279   "It's a very bloody transatlantic connection": In-person interview with Zandile Blay in New York City, 7/16/14.

280   "I don't think that diamonds have been marketed to black communities": BS to RB, 8/11/14.

282   the syndicate controlled only an estimated 65 percent of the

world's supply of diamonds: From GIA education materials, "Diamond & Diamond Grading: The Modern Diamond Market 3," © GIA 2002, updated in 2008.

283 However, a decade later, in July 2004, they made the decision to reconcile yet another American antitrust suit: Stephen Labaton, "De Beers Agrees to Guilty Plea to Re-Enter the U.S. Market," *New York Times*, July 10, 2004.

284 "There were some people who were not particularly keen, at least initially": Telephone interview with David Johnson (DJ), 8/26/14.

286 calls its diamonds "nature's most exquisite and magnificent prizes": From the De Beers commercial website, http://www .debeers.com.

286 In London in 2002, at the opening of De Beers's first brick-and-mortar store, protesters gathered: Sally Pook, " 'Save Bushmen' Protest Targets De Beers Store," *Telegraph*, November 22, 2002.

288 This company's purpose, Jolis explains, was to help De Beers get around President Roosevelt's antitrust laws: In-person interview with Jack Jolis in New York City, 8/11/14.

290 "makes De Beers very happy": Ibid.

291 In December 2011, Global Witness publicly withdrew from the Kimberley Process: "Why We Are Leaving the Kimberley Process—A Message from Global Witness Founding Director Charmian Gooch," December 5, 2011, http://www.globalwitness .org/library/why-we-are-leaving-kimberley-process-message -global-witness-founding-director-charmian-gooch.

292 "ask questions and only buy from companies that have clear policies": CG to RB, 7/11/14.

CHAPTER 16: THE INNOVATORS

297 Selva remembers that she and her teammates celebrated their little triumph: In-person interview with Rebecca Selva in New York City, 4/22/14.

297   "There really weren't that many people involved [in celebrity styling]": Ibid.

298   "They looked like they were borrowing jewelry": Ibid.

298   after Claire Danes wore a gold wrist cuff with a minimalist pink gown to the 2011 Golden Globes: In-person interview with Greg Kwiat (GK), CEO of Fred Leighton, in New York City, 4/22/14.

299   There, the statuesque redhead stunned in a simple black gown and an enormous, multistrand white diamond sautoir: Diamond Information Center, "Oscar Presenter Nicole Kidman Wears 1399 Carat Rough and Polished Diamond Sautoir Designed by L'Wren Scott to 80th Annual Academy Awards," PR Newswire, February 24, 2008, http://www .prnewswire.com/news-releases/oscar-presenter-nicole -kidman-wears-1399-carat-rough-and-polished-diamond -sautoir-designed-by-lwren-scott-to-80th-annual-academy -awards-57109802.html.

300   "I think it matters to customers at a certain level": GK to RB, 4/22/14.

300   "It's an important part of the strategy of our company": Ibid.

302   "In general people are already living together": In-person interview with Elizabeth Doyle (ED) in New York City, 1/16/14.

304   "They were once controlled by a family, or the family had a really strong presence": In-person interview with Andrea Buccellati in Milan, 3/10/14.

305   "that icy, blingy whiteness": ED to RB, 1/16/14.

305   "If you take an antique cut diamond today and compare it": In-person interview with Douglas Kazanjian in Los Angeles, 9/22/14.

306   "Do you want it to look beautiful in a laboratory": Ibid.

307   When Clarke arrived in Russia, he was in the market for a new kind of ink that was invisible to the naked eye: Carter Clarke's history comes from Tom Zoellner, *The Heartless Stone:*

A Journey Through the World of Diamonds, Deceit, and Desire (New York, Picador, 2006), 258–61.

308   "If consumers don't feel confident that when they buy something it is what they think it is": DJ to RB, 8/26/14.

309   Pure Grown Diamonds is a "snapshot," she says: In-person interview with Lisa Bissell (LB) in New York City, 8/7/14.

309   Once consumers understood that aboveground mines produce chemically identical diamonds: Frost & Sullivan, *The Diamond Growing Greenhouses: Grown Diamonds in the Gems & Jewelry Industry* (March 2014).

310   "It's important [to the millennials] that people are not having their hands and feet chopped off": LB to RB, 8/7/14.

315   In China, which in 2011 surpassed Japan as the world's second most voracious diamond consumer: Thomas Biesheuvel, "China Affair with Cheap Diamonds Heats Mass Market: Commodities," Bloomberg.com, May 1, 2013, http://www .bloomberg.com/news/2013-04-30/china-affair-with-cheap -diamonds-heats-mass-market-commodities.html.

315   the increased urbanization and meteoric expansion of the middle class: Bain & Company, *The Global Diamond Report 2013: Journey Through the Value Chain* (2013), 53.

315   "the global supply of rough diamonds is expected to peak in 2018": Ibid., 58.

316   "If it happens, then that's a great loss for everybody because that's part of the economy": LB to RB, 8/7/14.

316   "I think there's going to be a shortage of natural diamonds": LC to RB, 10/6/14.

Index

Abdul Hamid II, sultan of Turkey, 61, 68

"aboveground mines," 306, 309–11

Academy Awards (Oscars), 138, 139, 169–70, 293, 297–99

Adventure in Diamonds (formerly *Diamonds are Dangerous*), 139–40

advertising:
  in award shows, 138, 139, 168, 169–71, 293–99
  design in, 102–4, 113, 115, 239–40
  failures, 231–32
  four objectives in, 103
  free, 257, 260
  innovations in, 100–115, 153–54, 169–70, 299–306
  loaned diamonds in, 131, 150–51, 169–70

affaire des diamants, l' (the affair of the diamonds), 196

Affleck, Ben, 243

African Americans, 247–65, 278–80

Agra Diamond, 241–42

AIDS research, 210

AK1 pipe, 234, 243, 247

Alexandra, queen of England, 63

Alfred H. Smith and Co., 85

Alrosa, 281

American dream, xi-xiii

American Museum of Natural History, 89, 93, 146

Ames, Harriet Annenberg, 175, 178, 179

Amnesty International, 274

amputees, 272–74, 277, 280, 310

Amsterdam, 53, 59
  as diamond-cutting center, 41, 83, 183

Anglo American Corporation, 282, 284

Angola, 276
  diamond mining in, 268
  human rights violations in, 265, 267, 275, 279, 282, 291
  violent rebel campaign in, 267–71

Annenberg, Moses "Moe," 175

Annenberg, Walter, 175

anniversaries, 141, 187, 220, 223

Antwerp, Belgium:
  as diamond-cutting center, 41, 49, 83, 147, 183
  as diamond-trading capital, 237–38, 270, 274, 308

apartheid, 39, 180–81, 235, 281

aquamarine, 96, 119, 125

Arabo, Jacob "the Jeweler" (Yakov Arabov), xvi, 247–50, 255–57, 260–61, 263, 264, 278

Argyle mine, 235–45, 281

Arpels, Claude, 171, 172
Arpels, Louis and Julien, 86, 96
Ashton Joint Venture, 234–44
Asscher, Joseph, 88
Astor, John Jacob, 15
Astor, Mrs., 15, 17
Astor family, 3, 62, 163
Atherstone, W. Guybon, 23
Atlanta Constitution, 108
Atlantic, The, 219
auctions, 10–13, 15, 84, 130, 171, 174,
　　176–79, 241–42
Australia, 24, 276
　de Beers and, 234–40, 242–43,
　　245, 270, 281
　diamond industry in, 234–45,
　　316

baby boomers, 149, 153–54
BAFTA Awards, 168
baguettes, 141, 142, 171
Bain & Company, 314–15
Bakhmeteff, George, 75
Balanchine, George, 172
Balfour, Ian, 74
Baller, Ben (Ben Yang), 255, 261–64
Baltimore Sun, 201, 215
Bapst, Germain, 10–12, 314
Barnato, Barney (Barnett Isaacs),
　　34–38, 194
Barry, Stephen, 222
Bassey, Shirley, 278
Beau Sancy diamond, 49
"Been There, Done That," 258
Belgium:
　diamond cutting in, 41, 48–49,
　　90, 243
　diamond trade in, 268, 270, 273, 288
Bergman, Ingrid, 138
Berquem, Louis de (Lodewyk van
　　Berken), 48–50
Bessie (Wallis Simpson's aunt),
　　118–20, 126

Best Practice Principles, 285, 287
betrothal rings, 50–51
Beverly Hills Hotel, 170
Beyoncé, 301–2
bezel settings, 45, 257, 261
B.G. (Baby Gangsta), 258–59, 267
Bieber, Justin, 263
big-box stores, 244
Billionaire watch, xv–xvi
birthstones, tradition of, 42–43
Bissell, Lisa, 307–10, 312–13, 315–16
Biz Markie, 253, 255–56
Black, Starr, Frost & Gorham, 94,
　　137
Black Economic Empowerment, 281
Black Mafia Family, 261
Blay, Zandile, 279–80, 302
bling, use of term, 252, 259–60, 279
"Bling: Consequences and Reper-
　　cussions," 279
"Bling Bling," 267
*Bling'd: Blood, Diamonds, and Hip
　　Hop,* 279
Blink-182, 263
Blood and Sand, 138
Blood Diamond (film), xvi, 277–78,
　　281, 288
blood diamonds, use of term,
　　277–78, 311, 317
Blood Diamonds (Campbell), 277–78
Blue Book, 8, 16, 42, 52, 151
Blue Diamond of the Crown (French
　　Blue), 73–74
blue diamonds, 67, 73, 103, 241
　most famous, *see* Hope Diamond
Blue Nile, 303
Boers, 21, 23, 24, 27–28, 39
Bohlen und Halbach, Vera Krupp
　　von, 173
Borazon, 202
bort, 109–10, 203
Boston Globe, 171
Boston Saturday Evening Gazette, 43

Botswana, 276, 282, 284–86
mining in, 181, 281, 316
Boucheron, 5, 164
Boucheron, Gérard, 10–11
Bradley-Martin, Cornelia, 2–3, 5, 6, 13–16
Bradley-Martin ball, 1–6, 13–17, 41
Brady, James Buchanan "Diamond Jim," 229–31, 353
Brazil, as traditional source of diamonds, 4, 23, 26, 241
Breakfast at Tiffany's, 149–51
Briatore, Flavio, xvi
Bridgman, Percy, 199
brilliance, 49, 208
brilliants, 49–50, 61, 64, 69, 73, 74, 104, 122, 124, 167, 170–71, 174
Bristol Hotel, 66–69
Britain, British, 7, 10, 19–20, 47, 88, 141, 154, 157–58, 175, 197–98, 284
aristocracy of, 5, 117–34, 221–23
engagement ring marketing in, 189
filming in, 161–62
imperialism of, 32–34, 37, 38–39, 272
Rhodes in, 19–20, 31–32
South Africa under, 21, 24, 27, 82, 121, 180, 234
Brooklyn Eagle, 146
brown diamonds, 231–33, 237–40, 241, 244
see also champagne diamonds
Buccellati, Andrea, 304–5
Bulgari, 155, 162–68, 169–71, 296, 304
Bulgari, Constantino, 163–64
Bulgari, Giorgio, 163–64
Bulgari, Giovanni, 164–67, 171–72
Bulgari, Nicola, 164
Bulgari, Paolo, 164
Bulgari, Sotirio (Sotirios Voulgaris), 163–64

Buona Sera, Mrs. Campbell, 171
Burgundy, Charles the Bold, Duke of, 49, 50
Burke, Kareem "Biggs," 254
Burton, Richard, 162, 165–68, 172, 173–80

cabochon, defined, 130, 166–67
Calderón, Tego, 279
Cameroon, 290
Campbell, Greg, 277
Canada, 276, 281, 304, 315
canary diamonds, 10, 308
Cape Diamond (Eureka Diamond), 22–23
Cape Town, 20–21, 23–24, 122
Capote, Truman, 149
Cartier, 5, 59–62, 76, 85, 125, 131, 137, 142–43, 154, 160, 164, 166, 176–177–178, 209, 232, 264, 296, 304
challenges for, 94, 129
international expansion of, 62–64
in lawsuit against McLeans, 70–71, 75
whimsical animal designs of, 129–31
Cartier, Alfred, 59–60, 62, 63
Cartier, Elma Rumsey, 63–64
Cartier, Jacques, 60, 62, 63, 129
Cartier, Louis, 60, 62, 63, 64, 129
Cartier, Pierre, 60–75, 94, 95, 129, 132
Cartier Diamond, *see* Taylor-Burton Diamond
Cash Money Records, 254, 258
Cecil, George, 114
Central African Republic, 290
Central Kalahari Game Reserve, 286
Central Selling Organization (CSO), 121, 182–85, 242, 284
see also Diamond Trading Company (DTC)

Ceres Corporation, 227
chains, hip hop, 254–55, 257–58,
    262, 264, 280
Champagne Diamond Registry, 239
champagne diamonds, marketing
    of, 238–40, 243–44, 308, 316
Channing, Carol, 135–36
Charisse, Cyd, 140
Charles, Ernest "Monty," 184–85,
    285
Charles, prince of Wales, 221, 225
Charles VII, king of France, 47–49
Chatelain, Liz, 238–39, 316
Chicago Tribune, 45, 123, 128, 177, 207
children:
    exploited in rebel armies, 272,
    277–79
    indulged and pampered, 58–59,
    61, 66
China, 20, 268, 313, 315
Christie's auction (London), 241–42
Christmas:
    commercial promotions for,
    42–43, 105, 147, 187
    overstock merchandise after, 209
Chuck D, 279
Churchill, Randolph, 46
Churchill, Winston, 46, 100
Civil War, U.S., 9
claimholder system, 34
Clarke, Carter, 307–9
Cleopatra, 161–62, 165, 167–68
Clipse, 263
Coeur, Jacques, 48–50
Coiner, Charles T., 103, 115
Colbert, Claudette, 139
Cold War, 182, 216
Colesburg Kopje, 27
Colon, Pastor, Jr., 174–75
conflict diamonds, xvi, 267–92,
    299–300, 302, 310–11, 317
    as alleged hoax, 288–91
    De Beers response to, 281–91

use of term, 275, 277
Congress, U.S., 289
conical pavilion, 50
Connaissance des Pierres
    Précieuses (CPP), 189
consumerism:
    as economic stimulus, 3, 17
    elite shopping experience in,
    5–6, 59–61
    new markets for, 313–15
    social rebellion against, 178–80,
    188
"Controversy over Diamonds Made
    into Virtue by De Beers," 288
Conzinc Riotinto of Australia
    (CRA), 234–35
counterfeit diamonds, use of term,
    308
Court of Jewels travelling showcase,
    132–34
Craig, William, 230–31
Crawford, Joan, 144
Crenshaw, Ben, 233
Cribs, 265
Crystaline diamonds, 208
C. Tiffany and Son, 7
cubic zirconias (CZ), 226–28, 306,
    309, 311, 312
Cukor, George, 137
Cullinan diamond, 86–89, 92
cushion cut, 64, 96, 150, 241
customs regulations, diamonds
    in, 6, 61, 89, 237–38, 270–71,
    273–76, 290

Daily Boston Globe, 128
Dale, Dana, 139
Dalí, Salvador, 115
Danes, Claire, 298
Dapper Dan, 253
Darrow, Paul, 102–3
Darya-i Nur (Sea of Light) diamond,
    241

Dash, Damon, 254
David di Donatello Awards, 171
Davis, Sammy, Jr., 168, 251
DEAREST rings, 43–44, 51
Death Row Records, 254
De Beer, Johannes Nicolaas, 27, 28
De Beers Consolidated Mines Ltd.,
    37–40, 87, 201
De Beers diamond syndicate, 82,
    87–88, 110, 174
  Australia as threat to, 234–40,
    242–43, 245, 270, 281
  branding and retailing by,
    282–83, 285–88, 299
  in conflict diamond issue, 274,
    281–82, 288–89
  current active prospecting by, 316
  diamond acquisitions by, 269–70
  diamond hegemony of, 38, 40, 84,
    109–10, 179, 180–91, 193, 203,
    217, 245, 269, 274, 281–82,
    290, 308, 315
  exposé of, 216–21, 235
  geographic expansion of, 181–84,
    217
  leadership changes at, 99,
    282–84
  marketing and promotion strate-
    gies of, 89, 97, 99–109, 121–22,
    131, 138–39, 144–45, 178, 180,
    183, 187–91, 217–19, 222–25,
    227, 229, 231–33, 235, 240,
    260, 280, 282–88, 298–300,
    309, 314, 317, 318
  monopoly of, 40, 110, 180–82,
    200, 207, 218–19, 282, 283,
    288
  new international markets for,
    188–91
  privatization of, 282
  protesters against, 286
  retooled in response to social
    change, 280–91

  skepticism about motives of,
    288–91
  social change as challenge to,
    180–91, 193
  as sponsor at N.Y. World's Fair,
    94–95
  synthetic diamond made by, 203
  synthetic diamonds as threat to,
    193–211, 213–15, 228, 312
  in today's world, 283–88
De Beers Jewellery, 286–87
De Beers mine, 31–40
De Beers Rush (New Rush), 27–29
"Declaration of Faith, A" (C.
    Rhodes), 32
Democratic Republic of the Congo,
    275
department stores, 54–55
Derain, André, 115
Diamant (racehorse), 197
Diamants de la Couronne de France,
    Les (The Diamonds of the
    French Crown), 10–13
diamond blue color, of N. W. Ayer &
    Sons, 103
"Diamond Boom, The," 216
Diamond Corporation, 88
diamond cutting, cutters, xiv, 24,
    73, 88, 145, 174–75
  challenges of, 91, 109
  cut-grading system in, 305
  cut styles in, xv-xvi, 84, 305–6
  demonstration, 133
  expansion into America of, 41–42
  homogeneity in, 305
  innovation in, 48–50, 183–84,
    243–44, 305
  of Jonker diamond, 90–93
  of melee, 83, 183–84
  modern vs. antique, 305–6
  weight vs. beauty in, 73
  see also specific cuts
Diamond Dealers Club, 147, 225

Diamond Development Initiative,
    291
"diamond dream," 314
diamond dust, 48
diamond fashioning, 48–49, 69–70,
    79, 83, 88, 89, 95, 237, 243
    see also diamond cutting, cutters
Diamond Field, 26
"diamond horseshoe," 62, 84, 207
diamond industry:
    as benefit to indigenous people,
        281
    buying and selling in, *see* dia-
        mond trade
    changing social values as chal-
        lenge to, 173–91
    distribution in, xvi, 87, 184–85,
        242–43, 284–85
    downward trends in, 79–97,
        101–2
    effect of synthetics on, 306–13
    and fashion industry, 294–99
    future projections for, 314–16
    new international markets for,
        188–91, 313–15
    output and availability control in,
        35–40, 181, 236
    price control in, 35–40, 100, 179,
        181, 207, 219, 274, 283
    reform in, 285–86, 291–92
    in violent conflicts, *see* conflict
        diamonds
Diamond Information Center, 106
"diamond invention," 217–18
"Diamond Is Forever, A" (short
    film), 145
"Diamond Is Forever, A" (tagline),
    114–15, 136, 187, 223, 317
"Diamond Is For Now, A," 187–88
Diamond Mines of South Africa, The
    (Williams), 25
diamond mining:
    under Atlantic Ocean, 181

in Australia, 234–40
benefits to indigenous people
    from, 281, 287, 291
exposé of, 216–21
hardships of, xiii, 30, 87, 145, 217
indigenous populations exploited
    by, 21, 25, 30, 38–39, 100, 235,
    286, 289–90, 310, 313
projected depletion in, 315–16
thrill of, 29
World War I closing of, 82–83
see also specific mining regions
Diamond News, 26
Diamond Promotion Service, 106, 224
diamond rush, in South Africa,
    24–28
diamonds:
    affordable, 244, 310–11, 316
    changing cultural attitudes
        toward, 84–85, 178–80, 188,
        225, 260
    collections of, 110–11, 132, 154,
        160–61, 164–67, 230
    colored, 64, 69, 74, 133, 231–233,
        239–240, 241, 308; *see also spe-
        cific colors*
    democratization of, 53–55,
        145–46, 244–45
    in engagements, *see* engagement
        rings
    flawed, 312
    hardness of, xv, 22, 202, 204,
        208, 226, 318
    highest auction price for, 176
    increasing supply and availabil-
        ity of, 229–45
    as investment, *see* investment
        diamonds
    less expensive alternatives to,
        203–10, 226, 227, 309
    man-made, 312; *see also* lab-
        grown diamonds; synthetic
        diamonds

negative associations of, xiii-xiv,
    xvi, 65–66, 178–80, 274–75,
    277–78, 281–82, 299, 317
nicknames for, 259–60
paradoxical symbolism of, xiii-
    xiv, xvi, 299–300, 317–18
as preeminent gem, 44
rarity of, 40, 83, 315–16
scientific composition of, xiv, xv,
    193–94
significant, 10–13, 49, 71, 85,
    86–87, 132, 147
size factor in, 223–25
small, *see* melee
symbolic appeal of, xvi-xvii, 58,
    69, 78, 220, 292, 300, 314,
    317–18
uncut, *see* rough diamonds
vocabulary of, *see* 4Cs
see also specifically named gems
Diamonds: A Century of Spectacular
    Jewels (Proddow and Fasel),
    125
"Diamonds and the Draft," 108
"Diamonds Are a Girl's Best
    Friend," 136, 302
"Diamonds Are Forever," 278
"Diamonds from Sierra Leone,"
    278–79
diamond trade, xiii, 4
    auctions and estate sales in,
        84, 86, 130, 171, 174, 176–79,
        241–42
    conflict diamonds in, 267–92
    dealers in, 88, 93
    illicit practices in, 12, 35, 38, 110,
        178–79, 270, 274, 290, 312–13
    regulation changes in, 269–71,
        273–77, 302
    sales scams in, 220
    shopper's responsibility in, 292
    tenders in, 242–43
    see also sightholder system

Diamond Trading Company (DTC),
    284
Diana, princess of Wales, engage-
    ment ring of, 221–22, 225
DiCaprio, Leonardo, 277–78
Dignam, Dorothy "Miss Dig," 105–7,
    136, 140–41, 189
DiMaggio, Joe, 141
divorce, 117–18, 126, 128, 160
Dolce Vita, La, 162
Dominion Diamond Corporation,
    304
Donner, Vyvyan, 108
double-ring ceremony, 51
Doutrelon, Alfred, 12
Doyle, Elizabeth, 303, 305
Doyle & Doyle, 303
Dr. Dre, 253, 258
drug dealers, 252–53, 255, 261,
    278
dry sites, dry digging, 26–28
Dudley, Earl of, 24, 28
Dutch, 68, 87
    in South Africa, *see* Boers
Dutch East India Company, 20–21
Dyer, Elisha, Jr., 15

economic inequality, xvi, 2–4, 75,
    83, 97, 178–79, 222, 252–53,
    258, 262, 264, 316
economy:
    in challenges for diamond indus-
        try, 79–97, 213–14, 235–36
    downturns in, 2–4, 41, 107, 118,
        174, 178–79, 188, 213–14,
        235–36
    emerging markets in, 313–14
    rebounding of, 41, 53, 225–26,
        229, 236
Edinburgh, Philip Mountbatten,
    Duke of, 124
Edward VII, king of England, 60,
    63, 92

Edward VIII, king of England,
　　126–27
　　abdication of, 127–28
　　as duke of Windsor, 127–31
　　as prince of Wales, 119–21
Egypt, ancient, 51, 252, 262
Elizabeth, queen-mother of En-
　　gland, 121–22
Elizabeth II, queen of England, 207
　　coronation of, 141
　　marriage of, 124–25
　　as princess, 121–25
Ellis, Jabez Lewis, 7–8
Ellison, Robyn, 237, 240, 242
emerald cut, xv, 92, 142, 154, 160
emeralds, 42, 61, 70, 81, 85, 96, 112,
　　141, 164, 166–68, 179
emerging markets, 313–15
engagement rings, xvi, 43–46, 50,
　　58, 153, 179, 185, 219, 228, 244,
　　255
　　of celebrities, 139–44, 160, 166,
　　　167, 221–22, 243
　　cost of, 52
　　couples shopping for, 300, 302–3
　　drop in sales of, 188
　　expanding international market
　　　for, 188–91
　　future projections for, 315
　　history of, 50–52
　　marketing promotions for, 102–
　　　12, 145–49, 186–87, 223–25,
　　　231
　　for men, 229
　　present-day trends in, 300–301,
　　　310–11
　　third-finger placement for, 51,
　　　111, 148
　　"two months" salary guideline
　　　for, 224–25, 236, 304
　　uniqueness in, 303
environment, effect of mining on,
　　xiii, 235, 287, 291, 310, 313

Epstein, Edward Jay, 177, 216–21, 235
Essence, 280
"Eternal Gem, The," 144
eternity rings (diamond anniver-
　　sary rings), 141, 187, 223
ethnocentrism, 33, 38
Ettinger, Margaret "Maggie,"
　　138–40, 144
Eugénie, empress of France, 10, 13,
　　15–16, 63
Eureka Diamond (Cape Diamond),
　　22–23
Europe:
　　American elite in, 13, 59–61, 64,
　　　66, 132, 160, 206
　　as center of diamond culture, 5,
　　　9, 82, 313–14
　　engagement ring marketing in,
　　　189
European Union, 276, 290
Experiment 151, 200

Fabergé eggs, xvi, 62
"fair trade" diamonds, 291
"Famous Diamond Is Coming to
　　U.S.," 88
Farouk, king of Egypt, 93
Fasel, Marion, 122
fashion industry, 294–99
fashion jewelry, 240
faux diamonds, 203–10
Fellini, Federico, 162
fire, 208
　　defined, 49
Fisher, Eddie, 161, 165, 166
Fitzgerald, Zelda, 77
Five Stones (game), 22
Five Time Zone watch, 261
Flato, Paul, 93, 137–39, 169, 232,
　　242, 298
Flavor Flav, 251–52
Florentine diamond, 49
Ford, Gerald, 214

Forevermark, 286–87, 299

Fort Belvedere, 120

Fouquet, Jean, 49

4Cs (color, clarity, cut, and carat weight), 104, 112–13, 147

"Four Hundred," 17

France, 7–10, 21, 47–50, 73, 82, 129
auction of crown jewels in, 10–13, 15
engagement ring marketing in, 189, 191
Lemoine in, 194–99
political upheaval in, 8, 10, 68, 74

Frankel, Simon, 73

Fraser, Malcolm, 235

Fred Leighton, 287, 295–300

French Blue diamond, 73–74

"French Company," 36–37

French Revolution, 68, 74

FTC labeling guidelines, 311

Furness, Thelma, 119–21, 125

fur trade, 274

Gaborone, Botswana, 284–85

Galliano, John, 297–98

gangsta rap, 253, 258–59

gangsters, 206, 258, 260

Garde Meuble, 74

garnets, 324

Garrard, 10, 23, 124, 221–22

gay rights, 210

Gemesis, *see* Pure Grown Diamonds

Gemological Institute of America (GIA), 111–12, 145, 189, 239, 305, 312

gemologists, 111–12

General Electric, 64, 199–203, 211, 213, 226, 228, 236, 283, 306

Gentlemen Prefer Blondes, 135–37, 141

GEO, 216–18

George, Thomas, 259

George V, king of England, 74–75, 88, 99, 119

George VI, king of England, 121–22

Gerety, Frances, 113–14, 136, 287

Germany, 7, 59, 194, 198, 206, 216
engagement ring marketing in, 189, 191
Nazi, 96, 122, 129, 183, 214, 242
in South Africa, 82

Gilded Age, xiv, 1–18, 229–30, 231, 352

Gilfillan, Corinna, 274, 276, 292

gimmal rings, 51

"Girl's Best Friend," 258

glass, diamond capacity to cut, 22, 204, 208, 226

Glen Grey Act (1894), 39, 100

Global Witness, 267, 269, 271–76, 280, 288–89, 291–92

Golconda, India, 4, 132, 230, 241, 313, 316

gold, in jewelry design, 42, 45, 51, 96, 108, 164, 251, 257, 263

Golden Globes, 170, 294, 298

Goldman Sachs & Co., 52

gold mines, 31, 38, 57

gold rushes, 24

Gooch, Charmian, 288

Grace, princess of Monaco, engagement rings of, 141–44

Graham, Katherine, 204

Grant, Cary, 137

Great Depression, 118, 137
as challenge to diamond industry, 85–86, 93–94, 97, 99, 236

Great Table diamond, 241

Greer, S. C., 205

Griquas, 23, 24

"Guilt Trip" (Rapaport), 273

Habib, Selim, 65, 68, 73

Hall, H. Tracy, 200–202

Hamilton, Duchess of, 85

Hannaford, Gladys B. "Diamond Lady," 89, 146–48, 302

harems, 61, 67, 68
Hargitay, Mickey, 140
Harlem, 253
Harry Winston Inc., 85, 89, 91, 125,
    131, 137, 146, 155, 208, 245,
    283, 304
Hartford Courant, 133, 178, 228
"Have You Ever Tried to Sell a Dia-
    mond?" (Epstein), 219–20
Hayes, Isaac, 251
Hayworth, Rita, 138
Heart of the Matter, The: Sierra Leone,
    Diamonds and Human Security,
    273
heart shape cut, xv, 73
Hello! Nigeria, 279
Henrietta Award, 170
Hepburn, Audrey, 149–51, 162
Hepburn, Katharine, 137, 143
Herpers, Ferdinand, 45n
High Diamond Council, 275
High Society, 143–44
Hilton, Conrad, Jr. "Nicky," 160
Hinduism, 67–68, 72, 73
hip hop, 247–65
    conflict diamond issue in,
        278–80
    culture, 251–57
hippies, 179–80, 191, 225
Hirschfeld, Al, 207
Holiday, 137
Hollywood:
    child stars of, 90, 158–59
    diamond promotion using stars
        and films of, xii, xvi, 58, 68,
        75, 90, 134, 135–44, 149–55,
        157–72, 314
    see also specific individuals, films
        and companies
Hollywood Foreign Press Associa-
    tion, 170
Holocaust, xiv, 271
"hoodoo diamond," 70, 132

Hope, Henry Philip, 68, 72, 74
Hope, Henry Thomas, 72
Hope & Co., 68
Hope Diamond, 64–78, 86, 88, 147,
    151, 239, 241
    curse of, 65–72, 75–78, 132–33
    donated to Smithsonian, 154
    history of, 72–78
    refashioning of, 69–70
Hope Town, 22, 24
horseracing, 197
Hounsou, Djimon, 278
House of Jewels, 94–97
Houston Town and Country, 154
Hoving, Walter, 151–54, 171
Hoving Corporation, 151
human rights violations, 269,
    271–73, 277–81, 287, 289, 291
Huntington, Anna Hyatt, 85
Huntington, Arabella, 5–6, 84–85,
    96–97
Huntington, Archer, 85
Huntington, Collis, 5, 96–97
Hurwitz, Marty, 238–39
Hutton, Barbara, 163
"Hypnotize," 258

Ice Cube, 253
Idol's Eye diamond, 132
IF & Co., 261–64
"Improvement in Diamond Setting,"
    45n
India, 20, 189, 268, 276, 313–15
    as diamond-cutting center,
        243–44
    maharajas of, xiv, 63, 93, 230
    as traditional source of dia-
        monds, 4, 23, 26, 67, 73, 132,
        230, 241–42
industrial diamonds, 109–10, 182,
    233, 236, 239, 283
    synthetic, 199–203, 228, 307
insurance, 84, 88, 91, 175, 226

Internet retail, 303–4
investment diamonds, 4, 146–47,
    227, 235, 240
  threat from, 216–21
Iran, crown jewels of, 241
Isaacs, Barnett, *see* Barnato, Barney
Israel, 268, 276
  as diamond-cutting center, 147,
    183–84, 243
Italy, 7, 169
  motion picture industry in,
    162–64, 173
Ivory Coast, 275

J+J Jewelry Company, 249
Jackson, Michael, 263
Jacob (Arabo's cousin), 249
Jacob & Co., xv, 261
Jacobs, Erasmus Stephanus, 21–22,
    24
"Jacob" watch, 261
jade, 179, 225
"Jager" diamonds, 82, 208
Jagersfontein mine, 26, 82
Japan, engagement ring marketing
    in, 189–90, 236, 315
Jay Z, 254, 257–58
Jazz Age, 83, 93, 137, 260
jewelers:
  commercial competition and
    threat to, 54–55
  established, 5, 9
  homogeneity in, 305
  large conglomerate ownership of,
    304–5
  see also specific companies
Jewelers' Circular, 71, 230–31
jewelry:
  loaned to celebrities, 131, 150–51,
    169–70, 296–98
  vintage, 295–96, 303, 305, 313
jewelry design, 129–30, 151, 153, 166
  in rap music industry, 247–65

Jewels (ballet), 172
Jews, 174, 247
  in diamond industry, xiv, 34, 48,
    81, 99, 204
  persecution of, xiv, 183, 214, 271
  see also specific individuals
Johannesburg, 87, 88, 101, 114,
    180–81, 211
Johnson, David, 284, 308, 314
Jolis, Jack, 288–91
Jones, Jennifer, 169–70
Jonker, Gert, 87
Jonker, Johannes Jacobus, 87–88, 93
Jonker diamond (Number One),
    86–93, 97, 132, 146, 147, 154
Joseph Frankel's Sons, 73
Journey Through the Value Chain,
    314–15
Justice Department, U.S., 110, 283
J. Walter Thompson agency, 111, 190,
    236, 318

Kaplan, Lazare, 90–93
Kaplan, Leo, 90–92
Kazanjian, Douglas, 305–6, 313–14
Kazanjian Bros., 305
Kelly, Grace, *see* Grace, princess of
    Monaco
Kenmore, Robert, 176–77
Kennedy, John F., 141
Kennedy family, 163
Kenton Corporation, 177
Keystone, The, 54–55
Khan, Aga Mohammed, 46–47
Khoi (Hottentots), 21, 23, 25
Kidman, Nicole, 297–99
Kimberely, Australia, 234, 236
Kimberley, Lord, 234
Kimberley, South Africa (formerly
    Vooruitzigt), 31–34, 36, 38,
    234, 273
Kimberley Central Company, 34,
    36, 37

Kimberley mine "Big Hole," 27–30, 34–36, 46, 123, 181, 182, 194, 231

Kimberley Process (KP) Certification Scheme (2003), 275–77, 285
  skepticism about, 289–91, 311

kimberlite pipes, 30, 33, 34, 181, 200, 316

Kipling, Rudyard, 39

Kitty Foyle, 169

Knight, Marion Hugh "Suge," Jr., 254

Knot, The, 300

Koh-i-Noor "Mountain of Light" diamond, 10, 29, 46–47, 147

Kono district, Sierra Leone, 271–73, 277

kopjes (hills), 26, 27, 194

kopje-walloper (unprincipled diamond peddler), 35, 36

Krupp Diamond, 173–74

Kunz, George Frederick, 42

Kwiat, Greg, 300–301

lab-grown diamonds, 182, 211, 226, 306–13
  disclosure and transparency in, 308–9, 312
  mined vs., 306–7, 311–12, 316
  synthetic vs., 311–12
  tools for identifying, 312–13

Ladies' Home Journal, 232

Lauck, Gerold, 100–103, 113–14, 231

Lawrence of Arabia, 168

Lehman & Co., 52

Lemarchand, Peter, 129

Lemoine, Henri, "secret formula" diamond synthesis scam of, 194–99, 203

Lemoine Affair, The (Proust), 196

Leopold, king of Belgium, 66

Lesotho, 181, 217

Letseng mine, 181

Liberia, 271, 274

Life, 125, 232

light:
  optical dispersion of, xv, 45, 49, 50, 206, 226, 305–6
  rainbow, 49, 305

Lilienfeld, Gustav, 23–24

Lindbergh kidnapping, 78

Little Princess, The, 158

Lloyds of London, 84, 91, 93

Lollobrigida, Gina, 168–71

London, 9, 35, 51, 63, 88, 130, 157–59, 211, 269, 288
  De Beers store protested in, 288
  diamond market in, 23, 68, 121, 182, 184–85, 284
  social scene in, 118–21, 125

Loos, Anita, 135

Lopez, Jennifer, 243

Loren, Sophia, 140, 169

Los Angeles, 81, 112, 137–38, 158, 170, 255, 258, 261–65

Los Angeles Times, 6, 180, 221, 227, 228

Louis XV, king of France, 74

Louis XVI, king of France, 3, 15, 67–68, 73–74, 230, 252

Louisiana Purchase Exposition (St. Louis; 1904), 95

Luce, Clare Booth, 162

Luciano, Lucky, 258

Lusaka Protocol, 268

Lynn, Jeffrey, 139

McAllister, Ward, 17

McLean, Edward "Ned," 58–59, 64, 66, 69, 71, 76, 77–78

McLean, Emily, 76

McLean, Evalyn Walsh, 57–62, 64, 66–72, 75–78, 132

McLean, Jock, 78

McLean, John Roll, 58, 62

McLean, Ned (son), 76
McLean, Vinson, 66, 76–77
"Magic Stone, The," 144
Mail, U.S., diamonds sent in,
    88–89, 154
Makani, Johannes, 87–88
Malta fever, 161–62
Mansfield, Jayne, 140
Marange diamond fields, 291
Marcus, Herman, 94
Marcus & Co., 94
Margaret Rose, princess of England,
    122–23
Marie Antoinette, queen of France,
    9–10, 68, 74
Marie Claire, 189
Marie of Anjou, queen of France, 47
marketing, xvi, 180, 318
    celebrities in, xii, xvi, 58, 68,
        75, 90, 134, 135–44, 149–55,
        157–72, 243, 263, 293–99
    through educational and infor-
        mational campaigns, 103–7,
        111–12, 145–49, 189, 309
    grassroots, 280
    international expansion of,
        188–91
    of melee, 186–88
    present-day trends in, 293–318
    promotional strategies in, 86–93,
        100–115, 121–22, 131–34, 183,
        186–91, 207–8, 223–24, 229,
        231–33, 238–40, 243
    in response to social changes,
        281–88
    salesmanship strategies for,
        66–70, 73, 106, 168–69, 224,
        263–64
    successful, 148–49, 191
    see also advertising; *specific com-
        panies and campaigns*
marquise cut, xv, 92, 123, 141, 167
marriage:

    among baby boomers, 149
    effect of World War II on, 107–8
    proposals, 44–45, 50, 52, 80,
        300–301
    Shinto traditions in, 190
    social changes in, 187–88, 302–3
Martha (opera), 13–14
Martin, Bradley, 5, 14, 15
Mary, queen of England, 88, 119, 124
Mary of Burgundy, 50
Matabeleland (Zimbabwe), 38
Maximilian, Emperor, 50
May, Edna, 58
Mayer, Louis B., 158–59
Mazarins, 11–12
Mazer, Joseph, 83
M. Belmont VerStandig Inc., 205
Means, Gaston, 78
Meditations (Marcus Aurelius), 30
melee (small white diamonds), 83,
    183–88, 203, 222–23, 232
men:
    African-American, 250–57
    diamonds purchased by, 52,
        103–11, 136, 187, 203, 228, 233
    diamonds worn by, 46–47, 74,
        152, 229–33, 251
mercury poisoning, 49–50
Metro-Goldwyn-Mayer (MGM), 138,
    143, 158–59, 164, 169
Metropolitan Opera House, 13–14,
    62
Mikimoto pearls, 179, 311
millennials, 310–11
moissanite, 309, 312
Mondschein, Murray (Fred Leigh-
    ton), 295–96, 300
Monroe, Marilyn, 136, 141, 258
    biopic of, 293–95
Morgan, J. P., 15, 64, 100
Morgan family, 62
Morgan Memorial Hall of Gems, 89
Motion Picture Magazine, 140–41

MPLA (People's Movement for the
    Liberation of Angola), 268
Mughals, 46
museums, diamonds exhibited in,
    11, 89, 93, 137
music scene:
    diamonds and, 245–55, 278–79
    see also hip hop; specific performers
        and songs
Mussolini, Benito, 162
MVI Marketing, 238–39, 316
My Love Affair with Jewelry (Taylor),
    165
My Week with Marilyn, 294

Namibia, 181, 217, 269, 273, 276,
    281
Napoleanic Wars, 21
Napolean III, 10
Natal, 29
National Association of Jewelers, 42
National Family Opinion Inc.,
    148–49
National Jeweler, 223, 236
National Jewelers' Board of Trade,
    82–83
National Review, 289
National Velvet, 159
Native Lands Act (1913), 100
Nesline, Joe, 206
    wife of, 209
Netherlands, 20–21, 24, 41
New Netherlands Bank of New York,
    84, 85
Newport, R.I., 152
New Rush, see De Beers Rush
Newsday, 153, 154
Newsweek, 216
Newton, Isaac, 194
New York, N.Y., 54, 73, 120, 132,
    140–41, 144, 168, 247, 261,
    265, 309
    Cartier in, 64, 69, 131

    as diamond-cutting center, 42,
        83, 243
    diamond industry centered in,
        79–97, 100–102, 147, 171–72,
        243
    diamond promotions, 100–102,
        105–9
    jewelry district in, 6–7, 81–82,
        85–86, 230, 249
    social elite of, xiv, 1–18, 41, 137,
        230
New Yorker, The, 80, 81, 92, 103, 154,
    232
New York Herald, 143
New York State Retail Jewelers' Con-
    vention, 148
New York Times, 11–12, 16, 44, 65,
    70–71, 76, 88, 92, 122, 149, 152,
    171, 172, 175, 178, 207, 208, 259,
    275, 288
Nicholas II, tsar of Russia, 63
Nixon, Richard, 175, 206, 214
Notorious B.I.G., 253, 258
Number One, see Jonker diamond
Nur ul-Ain (Light of the Eye) dia-
    mond, 241
N. W. Ayer & Son, 100–115, 131, 136,
    137, 138–41, 144–45, 183,
    186–88, 189, 190, 217–19,
    223–25, 228, 229, 231, 236,
    240, 298, 301, 318

Oberon, Merle, 138, 139
Onassis, Aristotle, 176
Onassis, Jacqueline Bouvier Ken-
    nedy, 141, 143
O'Neal, Shaquille, 259
opals, 42, 43, 52, 297–98
open pit mines, 181, 183, 234
Operation Clean Sweep, 272, 277
Oppenheimer, Ernest, 99–101, 114,
    123, 180, 184, 235, 236, 282
Oppenheimer, Harry, 100–101, 114,

180–81, 184, 217, 224, 231, 235–36, 282

Oppenheimer, Mary, 123

Oppenheimer, Nicholas "Nicky," 282–84

Orange River, 22, 24

Orapa mine, 181

Order of the Golden Fleece, 74

O'Reilly, Jack, 22–23

Oxford University, C. Rhodes's preoccupation with, 19, 29, 31–33, 39

Paley, Babe, 163

Palmer, Bertha, 57–58

Palmer, Potter, 57

panic of 1893, 2–4, 41

Panthère, La (Cartier's signature image), 130, 171

Paramount, 149–50

Paris, 47, 70, 73–74, 168, 194–95, 261
    diamond culture in, 5, 9, 59–64, 129, 164, 171

Paris Exhibition (1878), 9

Parke-Bernet auction, 174, 176–79

Parsons, Louella O., 139

Partnership Africa Canada (PAC), 273, 276, 277, 291

pavé, 130, 240

pearls, 43, 61, 70, 85, 108, 124, 143, 174, 225
    cultured, 179, 311

pear cut, xv, 61, 85, 167, 174

Pelham-Clinton-Hope, Henry Francis, 68, 72–73, 75

Pell, Mrs. Herbert, 15, 16

pendants, hip hop, 254–55, 257–58

People, 210

Peregrina, La, 174

peridot, 42

Persia, 46–47

Pharrell, 263

Philadelphia, 54, 142
    as diamond cutting center, 42
    in marketing promotions, 100, 102, 105–6

Philadelphia Story, The, 143–44

Phillips, Frank, 112

Picasso, Pablo, 115
    art investment in, 215

Pierre Hotel, 171

pink diamonds, 241–43

platinum, in jewelry design, 42, 62–63, 69, 109, 129, 140, 141, 151, 257, 263

Plaza Hotel, 137

Pond's Extract Company, Pond's Cold Cream, 111

pop culture, diamonds in, 136, 172, 277–79

Porter, Cole, 138

Prada, Miuccia, 296–97

"Precious Stone Jewels" catalogue, 176

Premier Diamond Company, 82, 85

Premier mine, 86, 174

press:
    diamond discoveries covered in, 24
    diamond investment covered in, 215–16
    diamond promotion in, 64–65, 70, 72, 89, 91–93, 121–22, 125, 139, 142, 144, 154, 164, 175, 294–95
    elite covered in, 16, 76, 221
    sensationalist and tabloid, 126, 137, 144, 161–62, 165, 196–97

princess cut, xv

Proddow, Penny, 125

Prohibition, 83

Project Superpressure, 200–202, 211, 307

prospectors, 24–28, 34, 87, 93, 234

Proust, Marcel, 196

Public Enemy, 251–52, 279
Puff Daddy, 258
Pulitzer, Mrs. Joseph, 13
Pure Grown Diamonds (formerly
    Gemesis), 307–10, 312
Puritans, thimbles exchanged by, 51

Queen of Diamonds (E. W. McLean),
    61, 70

racism:
    blood diamonds and, 278
    in South Africa, 21, 38–39, 100,
        180–81, 235, 281
radio promotion, 145–46
Raekwon, 279
Rainier III, prince of Monaco,
    141–43
Rainsford, William, 2
Rapaport, Martin, 271–73, 277, 289,
    291
Rapaport Diamond Report, 271
Rawstone, Fleetwood, 27
Reader's Digest, 125
red-carpet, 294–99
REGARDS rings, 51
Regent Diamond, 11
Resolutions 1173 and 1176, 268–69
Revolutionary United Front (RUF),
    271–72, 274, 277–78
Reynolds, Debbie, 161
Reynolds, Evalyn McLean "Evie," 77
rhinestones, 16, 225
Rhodes, Cecil, 19–20, 24, 28–40, 55,
    99–100, 197, 217, 267, 300, 315
    Barnato bested by, 34–38, 194
    death of, 39
    diamond hegemony of, 38, 180,
        193, 282
    as prime minister of South
        Africa, 38–39
Rhodes, Herbert, 20, 24, 28–31
Rhodesia (now Zimbabwe), 38

Rhodes scholarship, 39
Rice, Glen, 259
Richemont group, 304
Rio Tinto Diamonds, 237, 242, 245,
    309
Rise and Fall of Diamonds, The: The
    Shattering of a Brilliant Illusion
    (Epstein), 218–20
Ritz-Carlton Hotel, 93, 133
Robinson, Bill "Bojangles," 90
Roc-A-Fella Records, 254–55, 278
Rockefeller, John D., Jr., 93
Rockefeller Center, 133
Rockefeller family, 62
Rogers, Ginger, 169
Rolexes, 257, 261
"Romance of Diamonds, The," 146
Roman Holiday, 162
Romanovs, 62
Rome, 161–68, 170, 171
    ancient, 50–51
Roosevelt, Eleanor, 44–45
Roosevelt, Franklin Delano, 44–45,
    109, 283, 288
Roosevelt, Theodore, 45
Roosevelt Hotel, Hollywood, 169,
    294
Rothschild, Nathan, 35, 36
rough diamonds, xiv–xv, 79, 89, 90,
    93, 95, 121, 147, 174, 183, 244,
    270, 283
    appearance of, 22, 25, 46, 89
    future projections for, 315
    largest, 86–87
    regulation changes for, 274–75
Rough Trade, A: The Role of Companies
    and Governments in the Angolan
    Conflict, 267, 269–71
royalty:
    American fascination with, 3,
        4–5, 58, 121, 142
    diamonds associated with, xiii,
        4–5, 46, 55, 60, 62–63, 67–68,

117–34, 189, 252, 314, 316
diamonds collected by, 9–11, 15,
    67, 73–75, 88, 93, 122, 128–29,
    230, 242
Hollywood stars as American
    alternative to, 134, 157
rappers inspired by, 252
see also specific monarchs and fami-
    lies
rubies, 42, 107–8, 142, 160, 164,
    166–67, 179
Rumsey, Elma, *see* Cartier, Elma
    Rumsey
Run-D.M.C., 251–52
rush rings, 52
Ruskin, John, 32–33
Russia, 63, 75, 78, 276, 281, 315
    lab-grown diamonds in, 307–8
    royalty of, xvi, 62

Sankoh, Foday, 277
sapphires, 74, 130, 139, 164, 221
sardonyx, 42
S. Bulgari, 163
Schenectady, N.Y., 200–201
Schlumberger, Jean, 151, 153, 296
science:
    in cutting, 305
    in diamond synthesis, 193–211,
        213, 306–13
Scientific American, 199
Scott, L'Wren, 299
Scott, Peter Paul, 252–53, 257
Sears, Richard Warren, 53–55
Sears, Roebuck & Co., 52–54, 101,
    207–8, 228, 303
Sebba, Anne, 129
security, 70, 74, 87, 90, 93, 152, 175,
    226–27
Selva, Rebecca, 296–98
Selznick, David O., 169–70
Seyne, La, 65, 73
Shakur, Tupac, 254

Shawty Lo, 255
Sherman Antitrust Act, 101, 110, 283
Shintoism, 190
Shipley, Robert M., 112, 145
Siberia, 182–84
Sierra Leone, 276
    human rights violations in, 265,
        271–73, 275, 277–79, 282, 291
sightholder system, 184–85, 242,
    284–86, 288
Sikhs, 47
silver, 9, 45n, 51, 62, 96, 152, 164
Simpson, Bessie Wallis Warfield,
        see Windsor, Wallis Simpson,
        duchess of
Simpson, Ernest, 118–20, 126–27
simulants, 203–10, 226, 309, 311–12
Singapore, 309–10
Singh, Ranjit, 47
"Single Ladies," 301–2
Six Voyages de Jean-Baptiste Tavernier,
    Les (Tavernier), 73, 241
Skinner, George D., 109
Skylark, 139
slaves, xvi, 21
Slick Rick, 251–52
Smillie, Ian, 291
Smith, Bevy, 251, 253, 255, 260,
    280
Smithsonian, 154
smuggling, 61, 275, 290
snake (Bulgari's signature image),
    171
sneakers, 262
Snoop Dogg, 253
social media, "ring selfies" on, 302
solitaires, 50, 108–10, 123, 141, 187,
    208, 250
Song of Bernadette, 170
Sorel, Agnès, 47–50
Sotheby's, 130, 171
Sothern, Sara, 157–59, 164–65
South Africa, 139, 196, 202–3, 276

South Africa (*cont.*)

diamond mines in, 4, 9, 19–40, 45, 86, 145, 146, 147, 174, 194, 217, 231, 234, 241, 269, 273, 315

first diamond discoveries in, 22–24

racism in, 21, 38–39, 100, 180–81, 235, 281

sanctions against, 180, 181

Windsor family's trip to, 121–25

World War I in, 82–83

see also De Beers diamond syndicate

South African Railways and Harbors, 122

Soviet Union, 247, 268

de Beers and, 182–84, 203, 223, 232, 270, 281

fall of, 281

synthesized diamonds in, 226

Spencer, Diana, *see* Diana, princess of Wales

Spencer, Win, 118

sports stars, 233, 259

Standard Steel Company, 230

Star of Africa, 92

Star of South Africa, 23–24, 28

Star of the East, 61, 64, 67, 70, 76, 132

status anxiety, 5, 6, 314

Stewart, Alexander Turney "A.T.," 6

stock market, 53, 201, 214

crash (1929), 85–86, 94

Strong, Herb, 200–201

strontium titanate, 206, 226

superstitions, xiv, 43, 46, 48, 52

debunking of, 133

see also Hope Diamond, curse of

"Suzy Says" column, 178

Switzerland, 206, 228, 275

synthetic diamonds, 193–211, 213, 225, 226–28, 236

lab-grown vs., 311

mined vs., 210–11, 213, 312

Tanzania (formerly Tanganyika), 181

Tavernier, Jean-Baptiste, 67–68, 70, 73, 75, 230, 243

Tavernier Blue diamond (*beau violet*), 67, 73, 241

Taylor, Elizabeth, xii, 157–68, 257

and Burton, 163, 165–68, 172, 173–80

diamond collection of, 160–61, 173–79

Todd's engagement ring for, 160, 166

Taylor, Francis, 157–58

Taylor-Burton Diamond, 174–80, 188, 221, 280

TechnoMarine, 261

Temple, Shirley, 90, 158

tenders, 242–43

That Uncertain Feeling, 138, 139

Tiffany, Charles Louis "King of Diamonds," 6–10, 13, 45, 55

Tiffany, Comfort, 7

Tiffany, Harriet Young, 6

Tiffany, Young & Ellis, 7–8

Tiffany & Co., 45, 52, 58, 85, 97, 101, 131, 137, 147, 239, 280, 283, 287

birthstones promoted by, 42

blue box of, 8

catalogue, 8, 16, 42, 52

display windows of, xv, 143, 147, 150

establishment of, 6–13, 16

expansion of, 154

Great Depression as challenge to, 94

leadership change and reinventing of, 151–54

less costly alternatives to, 52–54

naming of, 9

at N.Y. World's Fair, 96

prosperity of, 149–55
signature blue color of, 8
Tiffany and Young, 6–7
Tiffany Ball, 152
Tiffany Diamond, 10, 96
new setting for, 150–52
Tiffany Setting, 45, 50, 52, 201, 249, 303
Time, 232
Todd, Liza, 160
Todd, Mike (Avrom Hirsch Gold-bogen), 160, 161, 164, 166, 167, 174
Toison d'Or (Golden Fleece), 74
Tokyo, 261, 262
tourism, 123, 162–63
Toussaint, Jeanne, 129–30, 296
tremblant brooches, 166–67
Trowbridge, J. T., 104
Turkey, 62, 65, 67, 68
turquoise, 42
Tutankhamen, 252, 262
20th Century Fox, 90, 136, 161, 170
IIa Technologies, 309–10
2Pac, 254
Tyson, Mike, 253

Udall & Ballou, 94, 232
UNITA (National Union for the Total Independence of Angola), 268–71
United Hospital Fund, 133
United Nations, 180, 281
international diamond certifica-tion scheme of, 275–77
sanctions by, 268–71
United States:
as major consumer of diamonds, 41, 81, 82, 314
wealthy and social elite of, 1–18, 40, 57–78, 82, 84–85, 93, 111, 119, 141–42, 152, 163, 230, 252

Vaal River, 24, 26, 27, 30
Van Buren, Abigail "Dear Abby," 228
Van Cleef & Arpels, 86, 96–97, 125, 128, 131, 141, 143, 154, 155, 208, 296, 304
Vanderbilt family, 2, 3, 62
Vanguard, 122
Vanilla Ice, 253
van Niekerk, Schalk, 22–24
Van Stondeg family, 204
Ver Standig, Helen, 203–10, 215–16, 218, 226, 309
VerStandig, John, 206
VerStandig, Maurice Belmont "Mac," 204–9
Vever, 5, 60
Vibe, 280
Vietnam War, 174
V.I.P.s, The, 168
Virgin and Child (Fouquet), 49
Vogue, 107, 110–11, 125, 128, 137, 151, 152, 171, 187
Vooruitzigt farm, 27–28
Voulgaris, Sotirios, _see_ Bulgari, Sotirio

Waldorf hotel, 3, 14, 16–18, 41, 137
Wall, Paul, 272
Wall Street Journal, 207, 236, 289
Walsh, Thomas, 58–59, 61, 66
Wanamaker, John, 54–55
Wanamaker's department stores, 54–55, 58
Warhol, Andy, 172
Washington, D.C., 64, 154, 204–6, 291
social elite of, 57–58, 69–70, 76, 118, 132
Washington Post, 58, 71, 77, 108, 142, 204, 205, 215
Washington Times Herald, 78
watches, 257, 261

wealth:
  excesses of, xvi, 1–18, 57–78,
    178–80, 188, 257–59, 316
  new international sources of, 313
Webb, David, 179
wedding bands, 44, 51, 108, 139, 147,
  148, 228, 250, 300
Weinstein brothers, 293
Wellington, Madame (persona),
  203, 207–9, 216, 226
Wellington Jewels, 204, 207–11, 225,
  306
Wernher, Julius, 184–99
West, Kanye, 254–55, 257, 278–79
West, Mae, 144
Whalen, Grover, 95
Wharton School of business, 210
White, Barry, 251
Whitehouse, Mrs. Sheldon, 152
Whitney, Marylou, 227
Who's Afraid of Virginia Woolf?, 168
Wilding, Michael, 160, 180
Williams, Bryan "Birdman," 254
Williams, Gardner, 25
Williams, Hype, 279
Williams, Michelle, 293–95, 299
Willson, T. Edgar, 71–72
Windsor, Edward, duke of, see
  Edward VIII, king of England
Windsor, Wallis Simpson, duchess
  of, 117–21, 143
  jewelry collection of, 125–31
Windsor family, 121–26
  see also specific members
Windsor Jewels, 228
Winston, Edna, 80, 132–33
Winston, Harry, 79–93, 101, 154,
  249, 256, 298
  as "King of Diamonds," 84,
    174–75

promotional strategy of, 131–34,
  154, 169–70
women:
  diamonds purchased by, 233
  diamonds worn by, 4–5, 41–55,
    80, 84, 99–115, 136, 203, 223,
    228, 230, 232, 238–40
  promotional targeting of married
    and professional, 187–88
women's movement, 187–88
Woolworth family, 163
World Diamond Council, 274
World's Fair (New York;, 1939–
  1940), Jewelry Group at,
  94–97, 129
World War I, 118, 163
  as challenge to diamond indus-
    try, 82–83, 97, 99
World War II, 49, 109–11, 118, 121,
  122, 125, 138, 158, 163, 173, 182,
  205, 214, 242, 283, 299, 314
  American prosperity after, 131
  as challenge to diamond indus-
    try, 96–97, 128
  Japan after, 189–91
  marriage traditions affected by,
    106–8
Wu-Tang Clan, 272

Yang, Ben, see Baller, Ben
Yohe, May, 68, 75
Young, John Burnett, 6, 8, 9–10
Yugler, Al, 176
yuppies, 225
Yurman, David, 239

Zales, 185–86, 233
Zimbabwe (formerly Rhodesia), 38,
  291, 292

Photographic Sources

Cornelia Bradley-Martin: The Album File PR2, New-York Historical
    Society, © New-York Historical Society.
The 260-carat Billionaire watch: Courtesy of Jacob & Co.
Charles Lewis Tiffany: Irma and Paul Milstein Division of United
    States History, Local History and Genealogy, The New
    York Public Library. "Charles Louis Tiffany," The New
    York Public Library Digital Collections, 1899–1899, http://
    digitalcollections.nypl.org/items/54d6e150-86d6-0131-f846-
    58d385a7b928.
Maiden Lane: The Miriam and Ira D. Wallach Division of Art, Prints
    and Photographs: Print Collection, The New York Public
    Library. "South St. from Maiden Lane 1828," The New York
    Public Library Digital Collections, http://digitalcollections.
    nypl.org/items/510d47da-25f2-a3d9-e040-e00a18064a99.
Cecil Rhodes: The Miriam and Ira D. Wallach Division of Art,
    Prints and Photographs: Print Collection, The New York
    Public Library. "Portraits," The New York Public Library
    Digital Collections, http://digitalcollections.nypl.org/
    items/99a6ed7e-0d02-0e0d-e040-e00a18061e25.

The Premier mine in South Africa: Carpenter Collection, Prints & Photographs Division, Library of Congress, LOT 11356-39.

Eleanor Roosevelt's engagement ring: Courtesy of Franklin D. Roosevelt Presidential Library, Hyde Park, New York.

Jean Fouquet's *Virgin and Child Surrounded by Angels* [public domain], via Wikimedia Commons.

Evalyn Walsh McLean: Harris & Ewing collection, Prints & Photographs Division, Library of Congress, LC-H25-18444-FB.

The Hope Diamond: Courtesy of the Smithsonian Institution.

Pierre Cartier: Manuscripts and Archives Division, The New York Public Library. "Jewels—Pierre Cartier and woman with ring," The New York Public Library Digital Collections, 1935–1945, http://digitalcollections.nypl.org/items/5e66b3e8-99fd-d471-e040-e00a180654d7.

Shirley Temple: Keystone-France/Getty Images.

Marilyn Monroe: From *Gentlemen Prefer Blondes*, ©1953 Twentieth Century Fox. All rights reserved.

The House of Jewels: Manuscripts and Archives Division, The New York Public Library. "Jewels—House of Jewels Building." The New York Public Library Digital Collections, 1935–1945, http://digitalcollections.nypl.org/items/5e66b3e8-6f8a-d471-e040-e00a180654d7.

Pond's Cold Cream ad: Courtesy of Unilever.

Gladys B. Hannaford: Courtesy of Forsyth County Public Library Photograph Collection, Winston-Salem, North Carolina.

Audrey Hepburn: © Bettmann/CORBIS.

Elizabeth Taylor in the Taylor-Burton Diamond: © Bettmann/CORBIS.

Elizabeth Taylor's necklace from Bulgari: © Axel Koester/Corbis.

European diamond cutters: Matson (G. Eric and Edith) Collection, Prints & Photographs Division, Library of Congress, LC-M33-10542.

Madame Wellington: Courtesy of John VerStandig.

James Buchanan Brady: Manuscripts and Archives Division, The New York Public Library. "Amusements—Performers and Personalities—Bergen, Edgar—With Charlie McCarthy, Diamond Jim Brady and Lillian Russell." The New York Public Library Digital Collections, 1935–1945, http://digitalcollections.nypl.org/items/5e66b3e9-0ac2-d471-e040-e00a180654d7.

Diana, Princess of Wales: © Quadrillion/CORBIS.

Argyle colored diamonds: Courtesy of Rio Tinto Diamonds.

Argyle pink diamond: Courtesy of Rio Tinto Diamonds.

Jacob Arabo: Courtesy of Jacob & Co.

Forevermark diamond suite: Courtesy of Fred Leighton.

Ben Baller custom pendant: Courtesy of IF & Co.

Ben Baller Rolex: Courtesy of IF & Co.

Radiant-cut yellow diamond: Courtesy of Pure Grown Diamonds.

Pure Grown Diamonds' 3.04-carat stone: Courtesy of Pure Grown Diamonds.

About the Author

Rachelle Bergstein, the author of *Women from the Ankle Down*, works at a literary agency in New York. She lives with her husband and their son in Williamsburg, Brooklyn.